Artificial Intelligence and Bioethics: Perspectives

Editors

Luiz Vianna Sobrinho
Fiocruz and Estratégia Latino-Americana de Inteligência Artificial (ELA-IA), Cuiabá, Brazil

Leandro Modolo
Unicamp, Fiocruz and Estratégia Latino-Americana de Inteligência Artificial (ELA-IA), Cuiabá, Brazil

Maíra Araújo de Santana
Department of Biomedical Engineering, Universidade Federal de Pernambuco (UFPE), Brazil

Giselle Machado Magalhães Moreno
Department of Biomedical Engineering, Universidade Federal de Pernambuco (UFPE), Brazil

Fabiano Tonaco Borges
Department of Biomedical Engineering, Universidade Federal de Pernambuco (UFPE), Brazil

Wellington Pinheiro dos Santos
Department of Biomedical Engineering, Universidade Federal de Pernambuco (UFPE), Brazil

CRC Press
Taylor & Francis Group
Boca Raton London New York

CRC Press is an imprint of the
Taylor & Francis Group, an **informa** business
A SCIENCE PUBLISHERS BOOK

First edition published 2025
by CRC Press
2385 NW Executive Center Drive, Suite 320, Boca Raton FL 33431

and by CRC Press
4 Park Square, Milton Park, Abingdon, Oxon, OX14 4RN

CRC Press is an imprint of Taylor & Francis Group, LLC

Library of Congress Cataloging-in-Publication Data (applied for)

ISBN: 978-1-032-65737-0 (hbk)
ISBN: 978-1-032-69474-0 (pbk)
ISBN: 978-1-032-69477-1 (ebk)

DOI: 10.1201/9781032694771

Typeset in Times New Roman
by Prime Publishing Services

Acknowledgements

We would like to express our sincere gratitude to the Institute of Collective Health (ISC) and the Office of Culture, Extension, and Student Life (PROCEV) at the Federal University of Mato Grosso (UFMT) for their institutional support through the UFMT extension project titled "ELA-IA: Latin American Strategy for Artificial Intelligence" (Protocol: 210520231233011565). Furthermore, we extend our deep appreciation to the National Council for Scientific and Technological Development (CNPq) for awarding the Industrial Technological Development A (DTI-A) scholarship to Fabiano Tonaco Borges, process number: 380820/2024-2. The support of these institutions was instrumental in the development of this work.

Contents

A Bioethical Perspective for Human Flourishing in a Geometric Ethics

Fabiano Tonaco Borges and Carlos Eduardo Gomes Siqueira

The growing presence of artificial intelligence (AI) in various fields of knowledge and human practice raises complex ethical issues that require a deep and multidisciplinary examination. The book, *Artificial Intelligence and Bioethics: Perspectives*, proposes a comprehensive and diverse view of these issues, bringing together contributions from Latin American experts who offer unique insights from countries that are part of the so-called "Global Majority". It aims to explore not only the emerging ethical conflicts but also the possibilities of a structured and robust ethics to guide the development and application of AI.

As ethical and bioethical questions emerge with the advance of AI, fundamental questions arise: Who controls AI? Who decides on its use? Who holds the power to determine its applications and who is subject to its consequences? Are those who control AI using ethics and bioethics in its development? These fundamental questions highlight the need for a comprehensive ethical debate on AI control.

The bioethics of AI must address not only technical and operational aspects but also power dynamics and social accountability. Currently, AI is concentrated in the hands of very few corporations, in a world already experiencing extreme concentration of wealth and power. This concentration increases the risks of abuse and inequality, making the question of who holds the authority to establish AI usage rules and regulations even more urgent. The lack of transparency and concentration of power in the hands of a few businesses or governments can lead to decisions that do not reflect the collective well-being. Additionally, we are witnessing a race for AI, where large corporations and governments compete to achieve technological advantages, often without fully considering ethical and social impacts.

This book seeks to better understand concrete ethical and bioethical implications to contribute to a strong current of opinion capable of preventing the abuse of power and the exploitation of the poorest in our planet. Therefore, it is essential to establish controlling structures that promote inclusive and democratic participation, ensuring that decisions about AI are made fairly and equitably. These considerations are essential to ensure that AI serves the common good and respects human dignity.

While this approach addresses the material conditions and power dynamics in the governance of AI, it is also crucial to consider the philosophical foundations that guide ethical reasoning. By choosing the title of this preface, "Artificial Intelligence and the Geometric Ethics", we evoke the image of an ethics that, like geometry, is structured, logical, and systematic. This idea finds inspiration in the works of philosopher Baruch Spinoza, who, in his book *Ethics*[1], seeks to understand human behavior and emotions through a geometric and rational approach. For Spinoza, the laws governing nature and human conduct are immutable and can be understood in a clear and distinct manner, just like geometric theorems.

The metaphor of geometry is particularly relevant when considering the characteristics of AI. AI is built on algorithms and data that follow rigorous mathematical and logical rules. However, applying these technologies in complex human and social contexts demands an ethical reflection that should be equally scrupulous and systematic. It is not enough for algorithms to be technically correct; they must also be morally responsible.

Spinoza teaches us that the pursuit of truth and rational understanding can lead to a virtuous and fulfilling life. According to Spinoza, human flourishing (or '*conatus*') is achieved through the understanding of ourselves and our place in the world, leading to greater autonomy and empowerment. Applying this principle to AI development means seeking a deep and clear understanding of the ethical implications of our technological creations and ensuring that these technologies contribute to the well-being and flourishing of individuals and societies.

In the book *Ethics*, particularly in the section "Of the power of understanding or human freedom", he provides a profound framework for considering human flourishing in the context of artificial intelligence. Spinoza argues that true freedom arises from understanding the nature of reality and our place within it. This rational comprehension allows us to align our desires and actions with the natural order, leading to a state of flourishing or *beatitudo*. Thus, human blossoming is achieved through the power of understanding, where individuals live in harmony with their nature and the world around them guided by reason.

Applying Spinoza's insights to the realm of AI, we argue that AI systems should be designed and implemented in ways that enhance our understanding and support our rational capabilities. By fostering transparency, explainability, and ethical considerations in AI, we can ensure that it promotes human dignity, autonomy, and freedom. This alignment with Spinoza's vision of rational harmony underscores the importance of developing an AI that not only serves practical needs but also enriches human life, enabling individuals and societies to thrive in a technologically advanced world.

[1] Spinoza, B. (1677). *Ethica, ordine geometrico demonstrata* [Ethics, demonstrated in geometrical order] (R.H.M. Elwes, Trans.).

A "Geometric Ethics" for AI indicates building ethical principles that are universal, clear, and consistently applicable in various situations. Such precepts ideally ensure that AI systems are designed and implemented in ways that promote human flourishing, aligning with Spinoza's vision of a rational and harmonious life.

Clinical Bioethics, which addresses the ethical issues arising from advancements in biology and medicine, provides crucial insights for navigating the moral landscape of AI. It emphasizes principles such as autonomy, justice, beneficence, and non-maleficence, which are equally relevant in the context of AI. Integrating these bioethical precepts into AI ethics ensures that human dignity, rights, and well-being are prioritized in technological innovations.

It is essential to recognize the diverse perspectives that contribute to a comprehensive bioethical framework, because they make it more responsive to the global diversity of experiences and values. This book aims to incorporate perspectives from the "Global Majority", highlighting the unique ethical challenges and insights that arise from different socioeconomic, cultural, and political contexts.

This inclusive approach does not reject well-established values but rather complements them with voices and experiences that have traditionally been underrepresented. The integration of Latin American perspectives aligns with Spinoza's philosophy, which values the understanding of diverse viewpoints and the pursuit of collective human flourishing. Brazil, the largest country in Latin America, with its rich cultural diversity and complex socio-political landscape, adds unique insights into the ethical considerations of AI and bioethics, making those contributions particularly valuable.

Throughout the following pages, readers are invited to reflect on issues such as privacy and data protection, algorithmic responsibility and transparency, justice and equity in the distribution of AI benefits, and the need for regulation that keeps pace with rapid technological innovation. More than definitive answers, this book offers an invitation to ongoing dialogue and reflection, recognizing that ethics in the age of AI is a constantly evolving journey.

We begin with Karla Figueiredo and Raphael Augusto Teixeira de Aguiar, who tackle the urgent need to regulate AI. They examine ethical principles like transparency, fairness, and accountability, analyzing international and national regulatory initiatives. Their chapter highlights the necessity of aligning AI development with human values and societal needs while exploring different regulatory approaches across the globe.

Fermin Roland Schramm follows by exploring the evolution of bioethics in Brazil, characterizing it as both young and late. He provides a critical review of Brazilian bioethics, emphasizing its potential to address persistent and emerging health issues through a complex theory informed by decades of research and institutional activity.

Next, Wellington Pinheiro dos Santos, Maíra Araújo de Santana, Fabiano Tonaco Borges, Leandro Modolo Paschoalotte, and Giselle Machado Magalhães

Moreno discuss AI's transformative impact on health in the "Global Majority". They identify challenges such as technological dependence and brain drain, and propose strategic measures to achieve digital and health sovereignty, aiming for a resilient and just future where AI enhances public health and development.

Luiz Vianna Sobrinho follows by addressing ethical challenges posed by information technologies from a human rights perspective. He calls for a redefinition of human rights to safeguard human dignity and freedom in an era where AI blurs the boundaries between human and machine.

Priscilla Mesquita Buzzetti Matsushita and Cristyny L. Goncalves de Almeida compare data protection regulations in Brazil, Europe, the United States, and China. They highlight the importance of privacy, security, and transparency in health data processing, emphasizing the need for adherence to data protection laws to prevent algorithmic discrimination and ensure data self-determination.

Pedro Fraga Vianna delves into the political and theoretical dimensions of warfare drones, examining their cognitive aspects and strategic implications. He explores philosophical questions about drones as objects and their impact on modern wars, challenging traditional notions of combat and surveillance.

Leandro Modolo discusses the potential of mobile health (mHealth) to integrate health technologies through AI. He introduces the concept of "cybernetic integration in health", where mHealth applications automate feedback cycles between individual and population actions, enhancing personalized clinical care and public health outcomes.

Maíra Araújo de Santana, Giselle Machado Magalhães Moreno, and Wellington Pinheiro dos Santos next address the importance of transparency in AI through "Explainable Artificial Intelligence". They highlight the ethical and practical challenges posed by the lack of transparency in AI systems, especially in healthcare. The chapter reviews recent work on incorporating XAI techniques to make AI decisions and processes more understandable. By providing explanations for AI decisions, XAI enhances trust and safety in healthcare applications, from diagnostic support to personalized treatments and health management.

Finally, Ciro Augusto Floriani and Hernán Poblete Miranda analyze the ethical challenges related to using AI in elderly dependent care, including those in palliative care, in the chapter "Longevity and Exclusion: Ethical Challenges Related to the Use of Artificial Intelligence in Elderly Dependent Care". They discuss the potential risks and harms, operational difficulties, and the underlying logic of AI applications in a society marked by structural inequality and ageism. The chapter underscores the necessity of involving bioethicists in developing decision-making models to ensure that the protection—against socioeconomic injustices that force elderly people into lack of social and healthcare protection—remains central. This approach aims to enhance the quality of life and end-of-life care for elderly individuals within the framework of an expanding AI-driven care system.

We hope that this work will inspire professionals, academics, legislators, and all those interested in the future of AI to adopt an ethical approach that is both solid and flexible, rigorous, and compassionate. Just as geometry teaches us to see the world with precision and clarity, may Spinoza's "Geometric Ethics" and the foundational principles of bioethics guide us in building a future where technology and humanity coexist in a harmonious and just manner. Furthermore, let the inclusion of Latin American perspectives enhance our collective understanding and drive us towards a more equitable, democratic, and inclusive AI.

1 | Artificial Intelligence: Promises and Liabilities

Karla Figueiredo* **and Raphael Augusto Teixeira de Aguiar**

This chapter addresses the need and challenges of regulating Artificial Intelligence (AI). With the rapid development of AI, critical issues arise on how it should be regulated to mitigate its risks and maximize its benefits. The chapter explores the fundamental ethical principles for AI governance, including transparency, fairness, non-maleficence, accountability, privacy, and international cooperation. Additionally, it presents a detailed analysis of major regulatory initiatives and recommendations from international organizations, as well as governmental proposals in different countries up to April 2024. The text emphasizes the importance of regulation to prevent harm, promote equity, and ensure that AI development aligns with human values and societal needs. The chapter also discusses different regulatory approaches adopted by regions such as the European Union, the United States, China, among others, highlighting trends and future challenges in AI governance.

A Brief History of Artificial Intelligence (AI)

The Emergence of Artificial Intelligence

The concepts of Computer Science and Artificial Intelligence are closely and historically linked. The first concept was developed throughout the second half of the 19th century and the first half of the 20th century through contributions from scientists such as Charles Babbage (Randall, 1973), who envisioned the "Analytical Engine", a programmable mechanical machine; Ada Lovelace, a direct contributor to Babbage's work, recognized as the first programmer in history (Aiello, 2016); George Boole, a mathematician who developed Boolean algebra, fundamental for mathematical logic later applied in automated reasoning systems (Boole, 1854/1999); Alan Turing, who proposed the Turing machine as an abstract model of computation; and Alonzo Church, who

*Corresponding author: karlafigueiredo@ime.uerj.br

presented lambda calculus and formalized the concept of abstractly representing mathematical functions (Church, 1932; 1936). Their work laid the foundation for the mathematical theory of computation and the broad field of knowledge that followed.

Having worked for the British Army during World War II to help decode encrypted German messages, Alan Turing gained notoriety and became known as the "father of computing". It was Turing, the father of computing, who was responsible for some of the first ideas related to Artificial Intelligence: in his article "On Computable Numbers", published in 1936 by Princeton University (Turing, 1950), he introduced the concept of the Turing machine as a theoretical tool to understand what is computable, designed to formalize the calculation process and provide a precise definition of computability.

In the article "Computing Machinery and Intelligence", Turing (1950) proposed the famous "Turing Test", adapted from the imitation game, as a pragmatic way to evaluate a machine's intelligence. In the imitation game, a person would ask questions to two game participants, a man and a woman, and based on the typed responses, the evaluator would determine which answers were given by the man and which by the woman. Turing suggested that if a machine could deceive a human interrogator to the point that the interrogator could not distinguish whether they were conversing with a machine or another human, that machine would be considered intelligent.

Finally, in 1948, Turing wrote a paper called "Intelligent Machinery", in which he introduced some central concepts of what is now conventionally called Artificial Intelligence (AI), including the training of an artificial neural network. However, Turing never published this paper, and his ideas were reinvented and/or improved upon by other authors (Encyclopædia Britannica Inc., 2024).

The fact that the scientist known as the "father of computing" was also one of the first scientists to express ideas and concepts about AI shows how closely it is related to the theory of computation itself. As we will see next, the following decade (1950–1960) was very important for the conceptual and experimental development of the field, although naturally, the computational capacity at that time and the lack of significant publicly available data made its full flourishing very difficult.

In 1956, a group of researchers organized the Dartmouth Summer Research Project on Artificial Intelligence at Dartmouth University (Hanover, New Hampshire), which became a seminal event in the field. During the meeting, researchers John McCarthy, Marvin Minsky, Nathaniel Rochester, and Claude Shannon coined the term "Artificial Intelligence" and kickstarted research on the subject (Metz, 2021).

In 1958, Frank Rosenblatt, an American psychologist with a particular interest in learning more about the human mind through machine experiments, created the first known neural network, the Perceptron. Working for the U.S. Navy's Office of Naval Research, Rosenblatt used an IBM 704 computer to create a primitive neural network and demonstrate how the machine could learn to differentiate between cards marked on the left and those marked on the right after a series of attempts (Metz, 2021).

In 1960, Rosenblatt and his team built an upgrade of the Perceptron on a computer called "Mark I". This time, instead of differentiating the position of marks on the cards, the neural network could, through a supervised learning process, identify—albeit rudimentarily—the letter printed on a card (A, B, C, or D).

The process worked as follows: The Mark I had a modified camera. Instead of a film holder, it had a small square covered with 400 black dots, which were actually photosensors capable of detecting whether a particular point was black or white. Each time the machine analyzed a card with the letter "A", the photosensors examined each point on the card's surface. If at a specific coordinate on a card containing the letter 'A', the black point was more frequent than the white, a higher 'weight' was assigned to it, making it more important in the mathematical calculation aimed at recognizing the letter 'A'. Upon reading the next card, the machine could recognize the letter by performing a series of mathematical calculations and observing, for example, if most of the high-weight points were black (Metz, 2021).

When the machine began its analyses, its classifications were random and had a high error frequency. However, as the Mark I continued to analyze the cards and was informed whether it was right or wrong through human feedback (supervised learning), the machine could eventually learn from its mistakes, adjusting each of its mathematical calculations until they became infrequent.

During machine demonstrations, Rosenblatt would even disconnect some cables to "break" the connections between the motors acting as neurons. When reconnecting the wires, the machine would classify the letters randomly again, improving the recognition pattern as it analyzed a new volume of cards and received new human feedback for each. This way, Rosenblatt aimed to show that there was indeed learning by the computer (Metz, 2021).

At that time, Rosenblatt, who later became an associate professor of neurobiology and behavior at Cornell University, already envisioned machines capable of perceiving and recognizing the environment without human training or control in the future. Despite initial skepticism and the decline of enthusiasm, which, as explained later, led to the period known as the "AI winter", the principles of the Perceptron significantly influenced the modern AI revolution.

The "AI Winter"

It can be said that the period between the late 1950s and the end of the 1960s was, in a way, the "golden age" of AI. However, in 1969, Minksy and Papert, in their book *Perceptrons: An Introduction to Computational Geometry* (Minsky & Papert, 2017), pointed out significant limitations of the Perceptron neural network, such as its inability to solve nonlinearly separable problems.

For example, the concept of "exclusive or" is a logical operation that returns true if and only if one of its two arguments is true and the other is false. A point on the card exposed to the Perceptron, for example, can be black OR white, but not both at the same time. In practical terms, the Perceptron could not respond if the two colors were different from each other. As prosaic as the problem may seem, the Perceptron, a linear model, could not handle this concept as it represented a nonlinear function. This led to a period of stagnation for models based on neural networks, as the problem, well documented and analyzed by Minsky and Papert, was thought to be unsolvable.

Their work gradually spread skepticism about the concepts of "Artificial Intelligence" and "Neural Networks". In 1971, two years after its publication, a study commissioned by the British government on the progress of AI found a 'disappointment' with what had been achieved in 25 years of research, serving as a justification for cutting funding for nearly all projects in this field in the UK. Subsequently, cuts also occurred in the United States, starting the period known as the "AI Winter". During this time, it was very difficult (if not impossible) to obtain any funding for AI-related research, and the field experienced unprecedented disrepute: researchers who insisted on working on the subject were not well-regarded by the academic community (Metz, 2021).

Although Minsky was emphatic in his argument about the insolubility of the nonlinearity problem in neural networks, he was still an enthusiast of the field. He was one of the organizers of the seminal Dartmouth event in 1956. For this reason, he and other researchers focused on the field known as "Symbolic Artificial Intelligence", which sought to continue AI development in a more conventional way using algorithms written in programming languages.

In other words, symbolic AI researchers believed that instead of teaching a machine to react to external situations and contexts through training on a large volume of data, they could do it more conventionally by feeding it with encoded human knowledge ('symbols') and with conditional rule-based algorithms ("if... then...") to process that knowledge and thus guide decision-making. Symbolic AI was considered a more realistic, albeit less effective, alternative to facing the "Connectionist AI" represented by Rosenblatt's Perceptron (Metz, 2021).

The "Renaissance"

The "AI Winter" began to thaw in 1986 when three researchers proposed a feasible solution to the nonlinearity problem of neural networks: the backpropagation algorithm. This method is essential for the modern training of neural networks, especially in supervised learning applications.

A neural network is designed to learn patterns to make predictions or classifications based on input data. When it starts training, the weights assigned to each neuron are random because it has not yet been exposed to data that allows it to detect patterns. Therefore, its first outputs will also be random, without a significant accuracy rate. The goal of training is to continuously adjust the network's weights so that its predictions become increasingly accurate.

The backpropagation method is an algorithm used to calculate the gradient of a proposed loss function concerning the network's weights, considering a single input-output pair (used in supervised learning). By calculating the gradient one layer at a time, iterating from the final layer towards the initial layer, it avoids redundant calculations of intermediate terms in applying the chain rule (Rumelhart, Geoffrey, & Ronald, 1986).

In other words, backpropagation allows a neural network to compare the obtained output with the expected output after each new data exposure. This process is done in reverse: after calculating the error (difference between the obtained result and the actual value, or label), the output neurons pass it to the network's last layer, which adjusts its weights to minimize the error. This layer then passes error information to the previous layer, which also adjusts its weights, and so on. After all layers adjust—from the last to the first—the neural network is ready to analyze new data. Thus, with each cycle of input/propagation/output/backpropagation, the neural network learns if its result is 'hot', 'warm', or 'cold', and it makes new weight adjustments, minimizing the error a little more each iteration.

Although this method was proposed by Seppo Linnainmaa in 1976 (Linnainmaa, 1976) and standardized by Paul Werbos in 1982 (Werbos, 1982), it was researchers David Rumelhart, Geoffrey Hinton, and Ronald Williams who, with the publication of the article "Learning representations by back-propagating errors" in 1986, gave new momentum to research based on neural networks (Rumelhart et al., 1986).

Their work, by overcoming the limitations of the original Perceptron, allowed the architecture of multilayer neural networks known as Multi-Layer Perceptron (MLP): A model composed of neurons fully connected between layers that employ a nonlinear activation function in the hidden layers to distinguish data that cannot be separated linearly (Cybenko, 1989). The advent and rise of multilayer models gave rise to the concept of "deep learning".

Still in the 1980s, the work of other scientists such as John Hopfield and Geoffrey Hinton* must be mentioned. Both developed models that interconnected physics and neuroscience, with significant contributions that continue to have an impact on the development of new models for neural networks and machine learning.

Thus, the 1990s and 2000s witnessed a significant increase in the practical application of AI in all sectors, such as health, finance, industry, commerce, and others. Additionally, the accumulation of digital data became feasible with the reduction of storage costs, and its transmission was facilitated by the widespread dissemination of the Internet. The period known as the "AI Winter" thus came to an end.

During those decades, Convolutional Neural Networks (CNNs) were born, inspired by biological processes from the work of Kunihiko Fukushima (1980),

* The researchers were awarded the Nobel Prize for Physics in 2024.

considering that the connectivity between artificial neurons in neural networks reflects the organization observed in the visual cortex of animals. Individual cortical (biological) neurons respond to stimuli only in specific regions of the visual field, known as receptive fields. The receptive fields of different neurons partially overlap, thus covering the entire visual field.

Besides CNNs, Kunihiko Fukushima introduced the ReLU (rectified linear unit) activation function (1969), which became the most frequently used activation function in CNNs and deep networks in general. However, it was only in 1987, during the Neural Information Processing Workshop, that CNNs were officially introduced (Homma, Les & Robert, 1988).

Although CNNs were proposed in the 1980s, the advance in the 2000s was enabled by graphics processing units (GPUs), which significantly accelerate calculations performed during neural network training.

Additionally, Recurrent Neural Networks (RNNs) experienced significant advancements in the 1990s. The basic principle of these networks is the ability to send information back to themselves, thus maintaining a kind of "memory" of previous information (Sepp & Jürgen, 1997). This ability is useful for handling data sequences such as time series, text, speech, among others (Tealeb, 2018). In other words, an RNN does not look at a word one at a time but also considers how each word relates to the previous and subsequent ones, taking this history into account in its analysis.

However, there are two common problems of RNNs used to generate unsatisfactory results: the first was the "vanishing gradient problem", which occurred when the gradient values (quantities indicating how weights should be adjusted) became too close to zero (Hochreiter, 1991). This usually happens in very deep networks or with very long data sequences. As a result, weight updates became insignificant, delaying the learning process or even preventing the network from learning effectively. The second problem was the "exploding gradient," which occurred when gradient values increased exponentially, potentially causing instability in network training (Bengio, Simard, & Frasconi, 1994; Hochreiter, Bengio, Frasconi, & Schmidhuber, 2001).

To overcome these problems, researchers Hochreiter and Schmidhuber proposed in the 1990s a new RNN architecture known as Long Short-Term Memory (LSTM). LSTMs include components called "memory units" with special gates to regulate the flow of information (Sepp & Jürgen, 1997). These gates control what should be remembered or forgotten by the network, allowing LSTMs to retain useful information for a longer time and minimize the vanishing gradient problem. Thanks to this capability, LSTMs can learn long-term dependencies more efficiently than traditional RNNs, avoiding or minimizing the mentioned problems (Li & Wu, 2015).

LSTMs have become important tools for analyzing and generating data sequences, opening new possibilities for applications in speech recognition, natural language processing, and many other areas.

The 2010s and 2020s

The 2010s were marked by new architectures based on Deep Learning, with notable results in all areas. This led to significant advances in tasks such as image recognition, natural language processing, and games. It is noteworthy that all this was only possible due to the increase in computational power obtained with cloud computing and the use of GPUs for computationally intensive tasks that revolutionized the approach to AI and the volume of available data.

Additionally, in 2017, a group of researchers affiliated with Google published a seminal paper known as "Attention Is All You Need" (Vaswani et al., 2017) at the 31st Conference on Neural Information Processing Systems (NIPS, 2017). This work completely changed how machines understand and work with human language.

These researchers created a new way for machines to analyze text, giving attention to specific parts, and thus improving their efficiency. This understanding made it possible to evolve neural networks into an architecture known as "transformers", which proved crucial for the development of Large Language Models (LLMs) like GPT. By overcoming inherent difficulties in RNNs and their adaptations in understanding long or complex texts by focusing on specific parts of a text, the technology began to better understand the overall meaning and details of human language.

It can be said that the AI development boom experienced at the beginning of the 2020s is predominantly but not exclusively related to these two specific advances: the one obtained by the three researchers who published the paper on backpropagation in 1986 and the one obtained by the researchers who wrote the 2017 paper presenting the Transformer architecture.

In the current decade (2020), new architectures based on advances in Generative Adversarial Networks (GANs) (Goodfellow et al., 2014) and LLMs (Brown et al., 2020) have led to significant progress in the automatic generation of images and texts, with examples of applications made available by major companies like OpenAI, Microsoft, and Google.

Promises of Artificial Intelligence

Predictions and Qualified Decisions Based on Data Analysis

In general, it can be said that AI has the potential to benefit virtually any human activity that relies on data analysis to improve decision-making, encompassing virtually all fields of knowledge and human practice. The analysis of large volumes of data by Machine Learning algorithms makes it possible to detect patterns that would be difficult for human cognition to perceive (Cui, Xu, & Razzaq, 2022). The possibilities opened by this capability are many and include: a) the ability to predict events or outcomes that, depending on multiple determinants, would otherwise be imperceptible to humans; and b) obtaining social, behavioral, scientific, educational, clinical, commercial, logistical, and economic insights that can qualify decision-making in various areas, reducing errors and mitigating uncertainties (Cui et al., 2022; Rocha, Carvalho, & Aguiar, 2021; Savoia, 2023).

General Transformations in Efficiency and Productivity

The continuous development of AI technologies has been enabling the transformation of efficiency and productivity practices in multiple sectors. In industrial and commercial contexts, AI enables the automation of repetitive processes and the optimization of operations, resulting in cost reduction and increased production (Piccarozzi, Aquilani, & Gatti, 2018) as well as increased safety in case of dangerous or risky tasks (Yedla, Kakhki, & Jannesari, 2020; Domingues, Pereira, Moreira, & Prottes, 2020).

In the context of knowledge workers, AI applications can promote the automation of administrative and bureaucratic tasks. These functions, often characterized by their monotonous and unchallenging nature, are more efficiently performed by automated systems, freeing professionals to focus on activities that require creativity and critical thinking (Bullock, 2019). This reduction in operational load can not only enhance job satisfaction but also reduce the possibility of human error in repetitive activities.

Among the general benefits of AI concerning efficiency and productivity, the following stand out: a) process automation through the implementation of intelligent systems that assist or replace human labor in operational tasks, allowing teams to dedicate themselves to higher-value functions (Zhu, Shi, & Liu, 2022); and b) resource optimization through the use of optimization algorithms aimed at improving resource allocation and logistics, ensuring more effective management of materials and time (Ma et al., 2022).

The impacts of these transformations are extensive, ranging from reducing the need for manual and bureaucratic work to globally improving the quality and speed of services provided, establishing a new standard of operational efficiency in various industries (Pastor, Alcañiz-Lucas, García-Chamizo, & Platero-Horcajadas, 2019; Dwivedi et al., 2021). Next, we will see some benefits of AI in specific sectors of human activity:

Transformations in Specific Sectors

Economy and Finance

Artificial Intelligence (AI) is transforming the global economy in both developed countries and lower-middle-income economies (LMICs) (López Rico-Olarte, Blobe, & Hullin, 2022). We will first see AI applications in the financial market. Then, we will analyze its use to aid economic development in peripheral countries.

In the financial market, AI has been used to predict real estate values, assisting specialists in classification and prediction tasks (Um, Wu, & Zhang, 2014). Additionally, machine learning models are employed to enhance risk management, detect fraud, optimize algorithmic trading, and improve customer service (Addo, Guégan, & Hassani, 2018; Milojević & Redžepagić, 2021; Kaminskyi, Nehrey, & Zomchak, 2021). There are also models that analyze market trends and consumer behavior to create predictive models that improve investment decisions and reduce risks, while chatbots and virtual assistants drive personalization in customer service and banking operations (Dwivedi et al., 2021; Paliwal, Patel, Kandale, & Anute, 2021).

As an example of potential impacts on the economic development of LMICs, we can cite the use of machine learning techniques in predicting inflation in emerging economies such as Turkey (Ozgur & Akkoç, 2021). In general, AI can facilitate the management of economic systems and decision-making, improving resource allocation in scarcity scenarios (Wahl, Cossy-Gantner, Germann, & Schwalbe, 2018) and increasing efficiency and productivity to promote sustainable and inclusive economic development (Dwivedi et al., 2021; López Rico-Olarte, Blobe, & Hullin, 2022; Jiang, 2022). Additionally, AI enables in-depth analysis of large volumes of economic and market data, essential for attracting foreign direct investment and stimulating local development. These analyses can help countries identify market opportunities and minimize financial risks, promoting economic stability and long-term growth (Ozgur & Akkoç, 2021; Jiang, 2022).

Education and Learning

The use of AI in education is a rapidly growing field. Several studies highlight its potential impact on teaching and learning processes through personalized learning experiences, adaptive assessments, student performance analysis, and administrative support (Zawacki-Richter, Marín, Bond, & Gouverneur, 2019; Popenici & Kerr, 2017; Rios-Campo et al., 2023).

With the support of AI models, educators can build e-learning platforms and high-quality educational content tailored to the specific needs of different students (Mendoza, Hutajulu, & Lubis, 2022; Dziatkovskii, 2023; Schiff, 2020). They also benefit from chatbots that provide customized support to students and automate routine tasks (Su & Yang, 2023; Wang, Sun, & Zhou, 2023).

Moreover, educators use tools that facilitate learning assessment and student monitoring to help them understand their motivation and any problems or misconceptions that compromise their learning (Chaudhry & Kazim, 2021).

Students, in turn, can rely on resources and tools that encourage creativity and enable interactive, meaningful, personalized, autonomous, and problem-based learning experiences without necessarily being confined to the classroom, as these can occur in cyberspace, virtual reality devices, or as games (Mijwil, Aggarwal, Mutar, Mansour, & Singh, 2022; Al-Worafi, Hermansyah, Goh, & Ming, 2023; Mendoza et al., 2022; Neo et al., 2022). AI can also foster and enhance cooperation among students to improve learning outcomes (Kim & Lee, 2022).

Finally, AI-powered administrative tools can automate repetitive tasks to relieve the administrative/bureaucratic burden on teachers, optimize operations, and enhance the overall educational experience for students and educators (Zheng & Badarch, 2022; Ahmad, Alam, Rahmat, Mubarik, & Hyder, 2022; Rensfeldt & Rahm, 2022).

Agriculture and Livestock

The integration of AI in fields such as agriculture, food security, and animal health has proven promising. Machine Learning (ML) initiatives are being used in Indian agriculture for effective management of crops, water, soil, and livestock (Sumangala & Kini, 2022). ML models are used to automate industrial greenhouses in smart environments (Ferrández-Pastor, Alcañiz-Lucas, García-Chamizo, & Platero-Horcajadas, 2019). Other studies demonstrate its application in crop monitoring, yield prediction, soil analysis, and pest management. By using satellite image analysis, weather data, and soil samples, these systems provide insights to optimize production, reduce waste, and increase sustainability (Benos et al., 2021; Gavai, 2022).

In the food security sector, AI-based algorithms, especially Bayesian Network models, are being employed to predict future risks in the food supply chain, enabling preventive actions to avoid food insecurity situations (Gavai, 2022).

In animal health, AI is being used in predictive epidemiology, precision veterinary medicine, and the study of host-pathogen interactions to improve disease management and animal welfare (Ezanno, 2021). It is also contributing to increased livestock production and improved management practices through the optimization of productive processes and qualified decision-making (Dutt, Singh, Gaur, & Sahu, 2022).

In a broader scope, it can be concluded that AI is fostering sustainability and efficiency in food and agricultural systems, creating more resilient systems to meet the demands of the global population (Marvin et al., 2022).

Industry, Commerce, and Logistics

The incorporation of AI technologies in Industry 4.0 has been accelerating the automation and standardization of industrial processes, positively impacting operational efficiency and productivity (Piccarozzi et al., 2018; Zhu, Shi, & Liu, 2022). The improvement of electrical automation controls through AI algorithms has shown potential in enhancing control systems and operational efficiency (Ma et al., 2022), while the use of intelligent logistics management systems has enabled quick and appropriate decision-making regarding complex problems across supply chains (Wu & Ge, 2019; Shi, 2022; Zhu, Shi, & Liu, 2022). Additionally, AI is being used to optimize energy resources and improve waste management (Cioffi, Travaglioni, Piscitelli, Petrillo, & Felice, 2020).

In e-commerce, AI is being utilized in various forms, such as chatbots for direct communication with potential customers, recommendation engines for services and products, warehouse automation through software and robots, traceability of inputs, products, and shipments, visual search, and optimal pricing strategies (Xia, Yang, Huang, & Huang, 2019; Gururaj, 2021; Feng & Ye, 2021).

Urban Development and Smart City Planning

The possibilities of using AI to aid urban mobility are undoubtedly among the most popular and anticipated promises in the field. Services like Waze and Google Maps allow choosing between faster, shorter, or toll-free routes according to the driver's convenience. In the coming years, Intelligent and Connected Vehicles (ICVs) will enable both individual vehicle driving and the management of all traffic in an urban network to be centralized in a single intelligent instance capable of analyzing large volumes of traffic data, weather conditions, and historical patterns to make travel not only safer but also more efficient and optimized (Liu, Cheng, Zhang, Shen, & Mark, 2014; Liu, Hentenryck, & Zhao, 2019; Zhang, Wang, & Zhu, 2022). Intelligent and integrated traffic systems can, for example, facilitate the quick and safe movement of priority vehicles such as ambulances during emergencies (Priya, Paramasivam, Kanagaraj, & Paramasivan, 2023).

In addition to traffic issues, intelligent systems can provide greater protection against natural disasters in urban centers (Yiğitcanlar et al., 2020), promote sustainability through dynamic alternative energy capture and increased energy efficiency (Lim, Edelenbos, & Gianoli, 2019), and improve urban waste management (Fang et al., 2023).

Security and Justice

Artificial Intelligence (AI) is increasingly being used in public security, especially in assisting criminal investigations—through video image detection and data mining—and in predicting crimes to help make decisions about patrol routes and proactive policing (Du, Lin, Lv, Li, & Ding, 2022). Additionally, AI plays a particularly important role in combating cybercrime (Singh, Singh, Vege, & Singh, 2022; Dilek, Çakir, & Aydin, 2015), especially crimes that also use AI techniques and tools (King, Aggarwal, Taddeo, & Floridi, 2018).

AI is also experiencing growing use in the judiciary sector. Its capabilities can, over time, enhance human work or even replace it in tasks of lesser intellectual value, such as document searches. There is evidence that its use can improve judicial efficiency by facilitating deadline management through alerts and reducing the frequency of so-called occupational crimes committed by public employees in the exercise of their duty (Zhu & Zheng, 2021).

Health and Medicine

The application of ML techniques and tools in healthcare is becoming increasingly frequent, particularly in aiding the early diagnosis of complex diseases such as cancer and heart conditions through visual computing of imaging exams, slides, specimens, and graphical records (Agliari et al., 2020; Imran, Zaman, Imtiaz, Fayaz, & Gwak, 2021). Another area of interest is personalized healthcare through models capable of predicting the best treatments for a particular patient as well as their adherence to the proposed treatment (Imran, Zaman, Imtiaz, Fayaz, & Gwak, 2021). The field of genetics is also experiencing growing interest in AI due to the possibilities of grouping genetic subpopulations to provide personalized care (Lopez, Tucker, Salameh, & Tucker, 2018).

In the realm of healthcare service management and planning, AI is being used to optimize operational processes, predict demand, and identify areas for improvement in care delivery (Sahu & Kumar, 2022; Pappada, 2021). It is also possible to enhance clinical intervention strategies through reinforcement learning techniques (Gottesman et al., 2019).

In public health and surveillance, AI is being employed to predict disease outbreaks, monitor population health trends, and develop more effective intervention strategies (Hamamoto, 2022; Lee, Mohr, Street, & Nadkarni, 2019).

Finally, AI has been successfully used in the development of new drugs due to its ability to predict the three-dimensional structure of proteins from their amino acid sequences (Paul et al., 2021; Narayanan, Durga, & Nagalakshmi, 2022). Additionally, AI can benefit other phases of the drug production and distribution process by improving supply chain logistics and quality control. AI can also contribute to drug safety by predicting drug toxicity (Paul et al., 2021).

Ethical Challenges and Responsibilities

Because AI is a powerful tool, its increasing presence in various human activities naturally raises several concerns about its safe use, particularly regarding the following issues: ethical data use and respect for individuals' privacy, cybersecurity, biases and discrimination replicated by biased training data in models, ethical challenges in autonomous algorithms, transparency, and interpretability of predictions and decisions by ML. We will analyze these issues in more detail below.

Ethical Data Use and Respect for Individuals' Privacy

AI is a set of disruptive technologies with a significant economic impact that require a lot of data for proper model training. For this reason, it has become commonplace to say that "data is the oil of the 21st century" (Stach, 2023), and various privacy issues are frequently raised.

Currently, online life requires many user authentications and local cookie installations. These processes often collect data about people, such as Internet searches, browsing history, location, and even geographical movement information through smartphones (Lee, Chen, Chang, & Chen, 2018). Although most services claim to operate legally and use this information to feed personalized consumption recommendations for their customers, companies do not always clearly communicate to users what information will be collected and for what specific purposes. Furthermore, even when the collection of certain information and how it will be used is communicated to the user, it usually occurs through lengthy authorization terms at the time of application installation, which are commonly accepted without question—and likely without awareness of all the consequences of such acceptance (Hanlon & Jones, 2023).

For these reasons, a significant number of users around the globe unwittingly accept to give away personal data that will be used to train AI models that will accompany them practically all day. It is known, for example, that this mechanism plays an important role in establishing social media addiction, as it becomes adept at presenting posts on the user's timeline with a high probability of capturing their attention (Zakon, 2019).

Moreover, since these companies handle large volumes of data that may contain sensitive information, inadequate protection can lead to leaks that compromise individuals' privacy and security (Rao, P, Krishna, & Kumar, 2021; Nawaz et al., 2023). In several countries, large amounts of sensitive data are illegally traded, often as a result of illegal leaks.

We cannot fail to mention the existence of numerous traps aiming to capture data illegally, which are constantly sent to people in the form of emails, SMS messages, or messaging apps like WhatsApp. There are also countless fake websites that commonly mimic real pages or request the filling out of personal data.

Analyzing the health field specifically, it is not difficult to understand the consequences of the improper collection and use of sensitive personal data, particularly regarding individuals' health conditions (Chernyshev, Zeadally, & Baig, 2018). The unauthorized use of clinical data to train unethical AI models can cause severe damage, such as the rejection of job candidates and summary dismissal due to the prediction of existing or future diseases or even the possibility of imminent pregnancy. There may also be an unfair increase in monthly premiums, refusal, or even cancellation of health insurance plans in countries where legislation is permissive regarding users' contractual rights.

On the other hand, the use of AI, as suggested by the work of researchers Lui et al. (2020), also helps prevent the leakage of sensitive data based on AI technology.

Cybersecurity

The insecure development of AI-based models increases individual and collective risks, causing significant social impacts and mainly creating a negative image of AI and all the benefits it could bring if well utilized. Some of the risk points that could be mentioned are:

1. **Erroneous or Reckless Decisions:** AI-based algorithms can also make erroneous or harmful decisions due to flaws in the model's design and development, poor data quality or pre-processing, or a lack of understanding of the contexts in which they are applied (Fehr et al., 2022; Kelly, Karthikesalingam, Suleyman, Corrado, & King, 2019). These types of problems usually result from the poor operation of the professionals who develop the systems. Nonetheless, such issues can cause financial, physical, or emotional harm to users.

2. **Models Designed for Cybercrime:** Although AI has the potential to significantly contribute to solving various human problems, it can also be used explicitly to scale fraud and other types of crimes, especially through the internet (cybercrimes). Among the main types of cybercrimes enhanced by AI, we can mention the creation of fake texts, images, and 'deepfake' videos (a topic discussed in the next section) (Masood et al., 2022), and the use of AI-enhanced malware and ransomware disseminated through the "Internet of Things" (IoT) (Velasco, 2022).

Information Manipulation and Disinformation

AI algorithms can be used to create and disseminate false information, manipulating public opinions, influencing elections, and exacerbating social divisions. Many reports have explored this theme, but in all, it is clear that, for example, the misuse of generative model-based algorithms is a significant concern. On the other hand, numerous AI works are being developed precisely to combat false information (Berrondo-Otermin & Sarasa-Cabezuelo, 2023; Santos, 2023). The work of researchers Zhou, Zhang, Luo, Parker, & Choudhury (2023) and Trandabăț and Gifu (2023) sought to create a synthetic textual corpus, generating fake news to help create means to combat the spread of false textual information.

However, the problem is not only with fake news. Before the so-called "fake news", much had already been commented on about images that are being perfected by AI-based models and the danger this can represent. In some reports, there are descriptions of how to evaluate whether an image may be fake, for example, by observing hands and ears (these are parts of the body that AI has more difficulty reproducing). Additionally, reports indicate that this type of processing is not new: many people have already done some kind of tampering with images to improve or conceal something. Thus, it is important to note that generally, in cases of modification of audio, text, and images, there is a person deliberately manipulating the information. As mentioned earlier, inexperienced and poorly trained professionals can increase the risks of generating false information without intent.

Biases, Prejudice, and Discrimination

AI algorithms can reproduce and potentially amplify biases and discriminations inherent in the data used in the models' learning process, leading to unfair and discriminatory decisions in areas such as criminal justice and healthcare or even in recruitment processes. As indicated in Chen's (2023) study, for example, there are suggestive results that AI-based recruitment models have good potential to improve quality, increase efficiency, and reduce transactional work, which indicates bureaucratic and routine activities. However, "algorithmic bias" can result in discriminatory hiring practices based on gender, race, color, and personality traits.

It should be emphasized that one of the reasons for discrimination is related to data selection—meaning that discriminatory behavior in the real world is recorded in the data that feed AI models, resulting in equally discriminatory behavior. On the other hand, unlike human discrimination, this behavior by an AI model tends to be more reported in the press. The resulting repercussion can provide society with a reflection on the real origin of the problem, which lies in human behavior, not in the machine.

Ethical Challenges in Autonomous Algorithms

As described at the beginning of this chapter, ML-based algorithms require adjustments in their architectures and hyperparameters to achieve the best performance. Besides knowledge of the algorithms, it is also crucial to highlight that modeling begins with understanding the problem and determining the most effective approaches to solve it, as well as the available data knowledge. This has become a bottleneck for professionals who wish to use the benefits brought by the evolution of ML-based algorithms but lack sufficient or appropriate technical knowledge.

Thus, algorithms published under the name Automated Machine Learning (AutoML) have been the subject of investigations (Salehin, 2024) due to the development of methods and processes to make automatic learning available, enabling AI use by non-specialists while aiming to improve efficiency and accelerate automatic learning.

There are some areas where the use of autonomous algorithms raises specific concerns, such as autonomous vehicles and the military industry. The first area was even the subject of a well-known ethical experiment proposed by researchers from the Massachusetts Institute of Technology (MIT) in 2016: The Moral Machine Experiment—MME (Awad et al., 2018). Launched through an online platform, the experiment had a basic premise—a moving autonomous vehicle encounters an unexpected situation and consequently an unavoidable accident—and it needs to decide in seconds who to kill and who to save. Various scenarios were tested, and participants were invited to decide whom the system should sacrifice in emergency situations: Should the car occupants be spared over a pedestrian? Should humans be prioritized over animals? Should younger people be chosen over older ones? These are just some of the proposed scenarios, and significant differences were observed in responses from different countries, demonstrating that there is no natural consensus in solving an ethical dilemma that sooner or later will need to be truly addressed in autonomous car algorithms (Bigman, 2020).

The use of autonomous algorithms in military equipment such as drones also naturally raises concerns about responsibility for their actions, as it is unclear who would be held accountable for unforeseen autonomous actions (Metcalf, 2022). Human rights violation issues also emerge, such as the difficulty of these systems in distinguishing between combatants and non-combatants, a fundamental requirement of international humanitarian law (Wasilow & Thorpe, 2019).

Mitigating risks and insecurities is crucial for the continuation and survival of AI. Thus, incorporating ethical principles, good cybersecurity practices, appropriate regulation, and experienced and well-trained human oversight are points that should not be compromised. Additionally, transparency, accountability, and stakeholder participation are essential to ensure that the benefits of AI are maximized and the harms minimized.

AI Governance and Regulation

The renaissance of AI observed over recent years and the acceleration of its development from the late 2010s and 2020s brought inevitable questions: Does AI need to be regulated? If so, in what aspects? What should and should not be regulated? What should be the goals of this regulation?

While the first question has a more consensual answer—yes, AI should be regulated due to its disruptive potential and possible negative consequences arising from its development (Dempsey, McBride, & Bryson, 2021; Petit, 2017; Liu, 2020; Erdélyi & Goldsmith, 2018)—the remaining questions remain open. Next, we will analyze the most commonly raised issues regarding regulation and present the main existing mechanisms and proposals at the time of writing this chapter in April 2024. We will initially present the main regulatory initiatives or recommendations from international organizations and then the main government proposals approved or under discussion up to the mentioned period.

Why Regulate AI?

Several studies point to the need to regulate AI. Generally, the arguments can be grouped into six major ethical principles: transparency, fairness and justice, non-maleficence, accountability, privacy, and international cooperation (Erdélyi & Goldsmith, 2018; Jobin & Ienca, 2019).

Regarding transparency, it is expected that AI models be able to provide clear decisions to ensure their trustworthiness (Jobin & Ienca, 2019). Furthermore, this principle seeks to avoid information control by closed groups (Erdélyi & Goldsmith, 2018) and ensure dialogue, participation, and democracy (Jobin & Ienca, 2019).

In terms of fairness and justice, AI should be developed and implemented to ensure that there is no unfair discrimination or bias in its operations. To achieve this, it is necessary to ensure that models are trained with diverse and representative datasets to avoid perpetuating existing prejudices (Liu, 2020).

The principle of non-maleficence resonates similarly to the well-known Hippocratic aphorism of doing no harm (do not harm). It emphasizes the need to avoid harm and seriously consider any potential negative impacts of AI technologies (Jobin & Ienca, 2019). To this end, it must be ensured that AI systems are developed and used ethically, avoiding harm and externalities to individuals and society as a whole (Petit, 2017).

The principle of accountability implies that developers and operators of AI systems should be held responsible for actions and decisions made by these systems. For this reason, clear accountability and reporting mechanisms must be established in case of damages caused by AI system failures (Erdélyi & Goldsmith, 2018).

The principle of privacy has as its essential pillar the security of data as a requirement to ensure public trust in AI models. Thus, individuals' personal and sensitive information must be adequately protected against unauthorized access and misuse (Petit, 2017).

Finally, the sixth crucial ethical principle in AI regulation is collaboration and international cooperation. Due to its global impact and the consequences of its use, close collaboration between countries and organizations is essential to establish common ethical and regulatory standards to avoid regulatory discrepancies and promote the ethical development of AI on a global scale (Erdélyi & Goldsmith, 2018).

What Should Be Done?

In general, it is expected that:

1. Clear guidelines be established for the responsible collection and processing of personal data by AI systems (Jobin & Ienca, 2019);
2. The abuse of data collected by AI systems for unintended or harmful purposes be prohibited (Erdélyi & Goldsmith, 2018);
3. The right to informed consent be ensured before personal data is collected and processed by AI systems (Liu, 2020);
4. Sufficient transparency and explainability be ensured so that individuals can understand how their data is being used and make informed decisions about its use (Petit, 2017);
5. Measures be established to protect data collected by AI systems against unauthorized access, misuse, and breaches (Jobin & Ienca, 2019).

To make these regulations effective, it is expected that formal norms and guidelines be approved and disseminated, along with possible treaties, conventions, codes, and laws. Some authors also propose the creation of an international AI regulatory agency. This agency, based on interdisciplinary knowledge, could establish a unified framework for AI technology regulation and assist in developing AI policies worldwide cooperatively and harmoniously to address AI's ethical and legal challenges on an international scale (Erdélyi & Goldsmith, 2018).

Main Guidelines from International Organizations

United Nations (UN)

The UN has played a crucial role in defining international rules for AI. On March 11, 2024, the United Nations General Assembly approved Resolution A/78/L.49, entitled "Harnessing the Opportunities of Safe, Secure, and Trustworthy Artificial Intelligence Systems for Sustainable Development" (United Nations General

Assembly, 2024). The document highlights the potential contribution of safe and trustworthy AI systems to achieving the Sustainable Development Goals (SDGs). Recognizing risks such as deepening digital inequalities, discrimination, and human rights violations, it emphasizes the importance of a human-centered approach, ensuring that these systems are trustworthy, explainable, ethical, inclusive, and that they preserve people's privacy. Finally, the resolution stresses the urgency of reaching a global consensus on AI standards and promoting inclusive international cooperation to develop safeguards and interoperable standards.

In addition to the resolution, an important document produced by a task force called the Inter-Agency Working Group on Artificial Intelligence (IAWG-AI), the "United Nations System White Paper on AI Governance", was approved on April 16, 2024, by the High-Level Committee on Programmes (HLCP) (Inter-Agency Working Group on Artificial Intelligence, 2024). This document offers a comprehensive view of AI governance within the UN system, analyzing existing institutional models, relevant functions, and international normative frameworks applicable to AI governance. It recommends that the UN system use its resources to address the challenges of global AI governance, promoting scientific and political consensus building, monitoring, enforcement, and the development of technical and policy capacities.

World Health Organization (WHO)

In 2021, WHO published a document titled "Ethics and Governance of Artificial Intelligence for Health" (World Health Organization, 2021), addressing ethical and governance issues associated with AI use in health. It details essential ethical principles, analyzes challenges such as bias and privacy, and recommends governance practices to ensure AI promotes well-being without compromising human autonomy.

In 2023, another document was published: "Regulatory Considerations on Artificial Intelligence for Health" (World Health Organization, 2023), addressing regulatory considerations for the safe and effective use of AI. It also emphasizes the need for transparent documentation, rigorous validation of AI systems, and risk management throughout the product life cycle, as well as the importance of collaboration between developers, regulators, and health professionals to adapt AI technologies to clinical and regulatory needs, aiming to improve patient care and ensure compliance with ethical and legal standards.

Finally, giving special attention to Large Language Models (LLMs), which gained prominence from late 2022, WHO launched in 2024 the document "Ethics and Governance of Artificial Intelligence for Health: Guidance on Large Multi-Modal Models" (World Health Organization, 2024) as an update to the 2021 document to encompass this technology. The document discusses LLM uses in health in various areas, also highlighting their risks and proposing regulatory and governance measures for their ethical and responsible use, such as the design and development of general-purpose foundational models, measures developers can take to mitigate risks, and governmental actions in laws, policies, and public sector investments.

United Nations Children's Fund (UNICEF)

In 2021, UNICEF published the document "Policy Guidance on AI for Children" (UNICEF Office of Global Insight and Policy, 2021), discussing AI's impact on child development and proposing nine essential requirements for creating child-centered systems that respect their rights: (1) Support child development and well-being; (2) Ensure inclusion of and for children; (3) Prioritize equity and non-discrimination for children; (4) Protect children's data and privacy; (5) Ensure children's safety; (6) Provide transparency, explainability, and accountability for children; (7) Empower governments and companies with knowledge about AI and children's rights; (8) Prepare children for current and future AI developments; (9) Create an enabling environment for child-centered AI.

United Nations Educational, Scientific and Cultural Organization (UNESCO)

In 2021, UNESCO adopted the document "Recommendations on the Ethics of AI" (United Nations Educational, Scientific and Cultural Organization, 2022). The organization advocates for global norms that reinforce human dignity, human rights, and gender equality, as well as protect the environment, promote social and economic justice, and enable the building of peaceful, just, and interconnected societies. Additionally, the document proposes 10 areas of policy action, ranging from ethical impact assessment and ethical governance to data policies, development and international cooperation, environment and ecosystems, gender, culture, education and research, communication and information, economy and work, and health and social well-being.

In 2022, another document was launched: "Guidance for Generative AI in Education and Research" (United Nations Educational, Scientific and Cultural Organization, 2022), reinforcing the need for appropriate regulations to ensure that generative AI models like LLMs genuinely benefit educators, learners, and researchers. The publication discusses long-term implications for ethics, intellectual property, and the evolution of educational assessment methods, addressing the need for ethical regulations to protect privacy and ensure consensual content use, aiming to avoid homogenized responses that limit diversity of opinions. It also proposes human-centered regulation, emphasizing the validation of AI systems by educational institutions and the formulation of detailed policies. Finally, it suggests the creative use of AI to enrich curriculum *design and teaching and research practices.*

International Labour Organization (ILO)

In 2018, ILO published the document "Negotiating the Algorithm: Automation,

Artificial Intelligence, and Labour Protection" (De Stefano, 2018), written by Valerio De Stefano. The text discusses the impacts of automation and AI on work and existing labor protections, arguing that while much of the public debate focuses on the number of jobs that will be lost due to automation, it is also crucial to consider the quality of the jobs that will remain and how automation may alter labor relations. The document warns of the possibility that AI may be used to monitor and manage workers in ways that can be invasive and detrimental to human dignity and advocates that to avoid this problem, AI should be regulated based on human rights, primarily aiming to protect privacy and ensure that workers maintain some degree of control over their work environment.

In 2023, ILO published the document titled "Generative AI and Jobs: Policies to Manage the Transition" (Gmyrek, Berg, & Bescond, 2023). Written by Paweł Gmyrek, Janine Berg, and David Bescond, the publication seeks to minimize a frequent concern regarding the widespread use of generative AI worldwide: the widespread loss of jobs. The authors argue that there will not be an 'apocalypse' in the world of work but rather a significant transformation in task structures and occupational roles, with a greater likelihood of increasing work capacity than total replacement by AI. On the other hand, they warn that the effects of automation may vary between countries of different incomes and regions and that women may be particularly affected by automation in administrative support. Finally, the document advocates for the application of International Labour Standards to manage the transitions caused by AI advances and that, specifically regarding AI uses that enable work automation, policies on the subject should prioritize worker relocation and training rather than merely avoiding the loss of 'automatable' jobs.

Food and Agriculture Organization (FAO)

In 2021, FAO published the document "Digital Agriculture in Action: AI for Agriculture" (Elbehri & Chestnov, 2021), which systematizes various AI use cases in agriculture. Although its focus is on disseminating ongoing examples and technologies, the document also discusses ethical considerations regarding AI use in the sector, such as the displacement of human labor by automated solutions.

Organization for Economic Co-operation and Development (OECD)

OECD has been working since 2019 to establish AI principles and promote the interoperability of policies and regulatory approaches among member countries. In that year, an intergovernmental standard entitled "Recommendation of the Council on Artificial Intelligence" (OECD/LEGAL/0449) was launched (Organization for Economic Co-operation and Development [OECD], 2023). This document recognizes AI's potential for a positive societal transformation and warns of the challenges associated with this process, such as economic changes, competition, labor market transformations, and implications for democracy and human rights. To address these issues, the standard proposes five principles for governing trustworthy AI systems: inclusive growth, human-centered values and fairness, transparency and explainability, robustness, safety, and security, and accountability. It also makes five recommendations for national policies and international cooperation: (1) investing in AI research and development; (2) fostering a digital ecosystem for AI; (3) creating a favorable policy environment for AI; (4) building human capacity to handle labor market transformation; and (5) promoting international cooperation. The text was revised in 2023 to encompass new generative AI technologies.

OECD also supports the Global Partnership on Artificial Intelligence (GPAI) and efforts by the International Organization for Standardization (ISO) to develop AI-related standards.

International Monetary Fund (IMF)

In 2024, IMF published two studies exploring AI's economic effects and ongoing regulatory efforts. The first, titled "The Economic Impacts and the Regulation of AI: A Review of the Academic Literature and Policy Actions" (Comunale & Manera, 2024), covers aspects of growth, employment, productivity, and income inequality, as well as regulations focused on market competition, data privacy, copyrights, national security, ethical issues, and financial stability. The second study, titled "Gen-AI: Artificial Intelligence and the Future of Work" (Cazzaniga et al., 2024), focuses on AI's consequences for the labor market. It reports that the exact effects of AI on productivity and employment are still uncertain and that new technologies can both increase worker productivity and lead to significant job displacements. It finally recommends that advanced economies and developed emerging markets focus on updating regulatory frameworks and promoting sustainable AI development, highlighting the need for international cooperation to establish common standards and coordinate AI regulations across countries.

Government Proposals for AI Regulation

European Union

The European Union (EU) has been a leader in AI regulation. On April 21, 2021, the European Commission proposed the "Artificial Intelligence Act" (European Commission, 2021), which aims to create a harmonized framework for AI across the EU, addressing risks and ensuring safe and ethical AI use. The proposal categorizes AI systems into four risk levels: unacceptable risk, high risk, limited risk, and minimal risk, with corresponding regulatory requirements for each category.

Unacceptable risk AI systems are prohibited, including those that manipulate human behavior to the detriment of users or exploit vulnerabilities of specific groups. High-risk AI systems, such as those used in critical infrastructure, education, employment, and law enforcement, are subject to strict requirements, including risk assessments, documentation, transparency, human oversight, and robust data governance. Limited risk AI systems must meet transparency obligations, while minimal risk systems are encouraged to follow voluntary codes of conduct.

The AI Act also establishes the European Artificial Intelligence Board (EAIB) to oversee the implementation of the regulation, promote cooperation among member states, and ensure compliance.

In April 2024, the European Commission launched the "AI Liability Directive" (European Commission, 2024), which complements the AI Act by establishing a framework for liability and compensation in cases of harm caused by AI systems. The directive aims to ensure that victims can seek redress for damages and that developers and operators of AI systems are held accountable for their actions.

United States

In the United States, AI regulation has been fragmented, with various federal and state initiatives addressing different aspects of AI. However, several key developments have occurred in recent years:

- In February 2023, the White House Office of Science and Technology Policy (OSTP) released the "Blueprint for an AI Bill of Rights" (White House Office of Science and Technology Policy, 2023), outlining principles to guide the development and use of AI in the U.S. The document emphasizes the importance of privacy, fairness, transparency, accountability, and non-discrimination, providing recommendations for both public and private sector AI applications.
- The Federal Trade Commission (FTC) has issued guidelines for AI and algorithmic decision-making, emphasizing transparency, accountability, and fairness. The FTC has also taken enforcement actions against companies that violate consumer protection laws using AI technologies.
- In 2022, the National Institute of Standards and Technology (NIST) published the "AI Risk Management Framework" (NIST, 2022), providing a set of standards and best practices for managing AI risks. The framework aims to help organizations develop and deploy trustworthy AI systems by addressing issues such as bias, fairness, transparency, and security.
- Several states, including California, New York, and Illinois, have enacted or proposed legislation addressing AI use in specific sectors, such as hiring, healthcare, and law enforcement.

China

China has been actively developing its AI regulatory framework, aiming to become a global leader in AI by 2030. Key regulatory initiatives include:

- In July 2021, the Chinese government released the "New Generation Artificial Intelligence Development Plan" (State Council of China, 2021), outlining a comprehensive strategy for AI development and regulation. The plan emphasizes the importance of ethical AI, data security, and international cooperation.
- In January 2022, the Cyberspace Administration of China (CAC) issued the "Regulations on the Management of Algorithm Recommendation Services" (Cyberspace Administration of China, 2022), which aims to ensure transparency, fairness, and accountability in algorithmic decision-making. The regulations require companies to disclose the principles and mechanisms behind their algorithms, provide users with options to opt out of algorithmic recommendations, and ensure that algorithms do not promote harmful content or discriminatory practices.
- In March 2023, the National People's Congress (NPC) passed the "Personal Information Protection Law" (PIPL) (National People's Congress, 2023), which establishes comprehensive data privacy and protection requirements for AI systems. The law mandates that companies obtain user consent for data collection, implement robust data security measures, and ensure that AI systems do not infringe on individual rights or interests.

Japan

Japan has taken a collaborative approach to AI regulation, involving various stakeholders in developing guidelines and policies. Key initiatives include:

- In 2019, the Japanese government released the "AI Utilization Principles" (Japanese Government, 2019), which outline ethical guidelines for AI development and use. The principles emphasize respect for human dignity, privacy protection, transparency, and accountability.
- In 2021, the Ministry of Economy, Trade, and Industry (METI) published the "AI Governance Guidelines" (METI, 2021), providing a framework for managing AI risks and ensuring ethical AI use. The guidelines address issues such as data quality, bias, transparency, and accountability, and recommend measures for promoting AI literacy and public engagement.
- In 2023, the Japanese government launched the "AI R&D and Implementation Strategy" (Japanese Government, 2023), which aims to promote AI innovation while ensuring ethical and responsible AI use. The strategy includes initiatives for developing AI talent, supporting AI research and development, and fostering international collaboration on AI governance.

Canada

Canada has been actively developing its AI regulatory framework, focusing on ethical AI use and human rights protection. Key initiatives include:

- In 2021, the Canadian government released the "Directive on Automated Decision-Making" (Treasury Board of Canada Secretariat, 2021), which establishes guidelines for the responsible use of AI in federal government operations. The directive emphasizes the importance of transparency, accountability, and fairness, and requires departments to conduct impact assessments and obtain external audits for high-impact AI systems.

- In 2022, the Canadian government launched the "Pan-Canadian Artificial Intelligence Strategy" (Canadian Institute for Advanced Research, 2022), which aims to promote AI innovation while ensuring ethical and responsible AI use. The strategy includes initiatives for supporting AI research and development, fostering AI talent, and promoting public engagement on AI governance.
- In 2023, the Office of the Privacy Commissioner of Canada (OPC) published the "Guidance on AI and Privacy" (Office of the Privacy Commissioner of Canada, 2023), which provides recommendations for ensuring data privacy and protection in AI systems. The guidance emphasizes the importance of obtaining user consent, implementing robust data security measures, and ensuring that AI systems do not infringe on individual rights or interests.

Future Challenges in AI Governance

While significant progress has been made in AI governance and regulation, several challenges remain. These challenges include:

Ensuring Global Cooperation and Harmonization

AI development and deployment are global in nature, and its impacts transcend national borders. Ensuring global cooperation and harmonization of AI regulations is essential to address cross-border challenges, avoid regulatory fragmentation, and promote the responsible development and use of AI technologies. International organizations, such as the United Nations and the OECD, play a crucial role in facilitating dialogue and cooperation among countries to establish common standards and best practices.

Addressing Bias and Discrimination

One of the most pressing challenges in AI governance is addressing bias and discrimination in AI systems. Ensuring that AI systems are trained on diverse and representative datasets and that they undergo rigorous testing for biases is essential to prevent unfair and discriminatory outcomes. Developing standardized methods for bias detection and mitigation, as well as establishing clear guidelines for fairness and accountability, are critical steps in addressing this challenge.

Enhancing Transparency and Explainability

Transparency and explainability are fundamental principles of trustworthy AI. Ensuring that AI systems provide clear and understandable explanations for their decisions is essential for building trust and accountability. Developing standardized frameworks and tools for explainability and promoting transparency in AI system development and deployment are key to enhancing the transparency and explainability of AI systems.

Protecting Privacy and Data Security

AI systems rely on vast amounts of data, raising significant privacy and data security concerns. Ensuring that AI systems comply with data protection regulations, such as the General Data Protection Regulation (GDPR) in the European Union, and implementing robust data security measures are essential to protect individuals' privacy and prevent data breaches. Developing standardized guidelines for data protection and promoting best practices for data security are critical to addressing these concerns.

Ensuring Accountability and Liability

As AI systems become more autonomous and capable of making decisions with significant impacts, ensuring accountability and liability for their actions is crucial. Establishing clear frameworks for accountability and liability, including mechanisms for redress and compensation in cases of harm, is essential to ensure that developers and operators of AI systems are held responsible for their actions. Promoting transparency in decision-making processes and ensuring that AI systems are subject to human oversight are key to ensuring accountability.

Promoting Ethical AI Development and Use

Promoting ethical AI development and use is essential to ensure that AI technologies align with human values and societal needs. Establishing clear ethical guidelines and principles, such as those outlined by international organizations and government initiatives, is crucial to guide the development and deployment of AI systems. Promoting public engagement and education on AI ethics and governance and fostering a culture of ethical AI development within the AI research and development community are key steps in promoting ethical AI development and use.

Addressing the Impact on Employment and Workforce

The widespread adoption of AI technologies has significant implications for employment and the workforce. Ensuring that workers are equipped with the necessary skills and knowledge to adapt to the changing job landscape and that there are mechanisms in place to support workers affected by automation and AI is crucial. Promoting policies that support worker retraining and reskilling, as well as fostering a culture of lifelong learning, are essential to address the impact of AI on employment and the workforce.

Fostering Innovation and Ensuring Safety

Balancing the need for innovation with the imperative to ensure safety is a key challenge in AI governance. Ensuring that regulatory frameworks promote innovation while ensuring that AI systems are safe, reliable, and aligned with ethical principles is essential. Developing flexible and adaptive regulatory approaches that can keep pace with rapid technological advancements and promoting a culture of safety and responsibility within the AI research and development community are key steps to fostering innovation while ensuring safety.

Addressing Ethical Challenges in Autonomous Systems

As AI systems become more autonomous, ethical challenges related to decision-making, responsibility, and accountability become increasingly complex. Ensuring that autonomous systems, such as self-driving cars and autonomous weapons, are developed and deployed in ways that align with ethical principles and human values is crucial. Establishing clear ethical guidelines and frameworks for the development and use of autonomous systems and promoting public engagement and dialogue on the ethical implications of autonomous AI are key steps in addressing these challenges.

Promoting Inclusive and Equitable AI

Ensuring that AI technologies are inclusive and equitable and that they benefit all segments of society is essential. Promoting diversity and inclusion in AI research and development and ensuring that AI systems are designed and deployed in ways that address the needs of marginalized and underserved communities are crucial. Developing policies and initiatives that promote equitable access to AI technologies and addressing systemic biases and inequalities in AI development and deployment are key steps in promoting inclusive and equitable AI.

Conclusion

The rapid development of AI presents unprecedented opportunities and challenges for societies around the world. The promises of AI include improved efficiency and productivity across various sectors, enhanced decision-making through data analysis, and transformative impacts on fields such as healthcare, education, agriculture, industry, commerce, and public services. However, these benefits come with significant ethical, legal, and social challenges that must be addressed to ensure that AI technologies are developed and used responsibly.

The ethical challenges associated with AI include issues related to data privacy, cybersecurity, bias and discrimination, transparency and explainability, and accountability and liability. Addressing these challenges requires a comprehensive and multi-faceted approach that includes developing and implementing ethical guidelines and standards, promoting transparency and accountability in AI systems, ensuring robust data protection and cybersecurity measures, and fostering international cooperation and collaboration.

AI governance and regulation are essential to ensure that AI technologies are aligned with human values and societal needs. Several international organizations, including the United Nations, WHO, UNICEF, UNESCO, ILO, FAO, OECD, and IMF, have proposed guidelines and recommendations for AI governance and regulation. These guidelines emphasize principles such as transparency, fairness, non-maleficence, accountability, privacy, and international cooperation.

Government initiatives in regions such as the European Union, the United States, China, Japan, and Canada have also focused on developing regulatory frameworks to address the risks and challenges associated with AI. These initiatives aim to ensure the safe, ethical, and responsible development and use of AI technologies, while promoting innovation and economic growth.

Future challenges in AI governance include ensuring global cooperation and harmonization, addressing bias and discrimination, enhancing transparency and explainability, protecting privacy and data security, ensuring accountability and liability, promoting ethical AI development and use, addressing the impact on employment and the workforce, fostering innovation while ensuring safety, addressing ethical challenges in autonomous systems, and promoting inclusive and equitable AI.

As AI continues to evolve and its impact on society becomes more profound, it is crucial to maintain a human-centered approach to AI development and governance. This approach should prioritize the well-being, rights, and interests of individuals and communities, ensuring that AI technologies contribute to the common good and do not exacerbate existing inequalities or create new forms of harm.

In conclusion, the responsible development and use of AI require a collaborative effort involving governments, international organizations, industry, academia, and civil society. By working together to address the ethical, legal, and social challenges associated with AI, we can harness its potential to create a more just, equitable, and sustainable future for all.

References

Addo, P.M., Guégan, D., & Hassani, B. (2018). Credit risk analysis using machine and deep learning models. *Risks*, 6(2), 38.

Agliari, E., Barra, A., Burioni, R., Di Biasio, A., Guerra, F., & Tavani, F. (2020). Deep learning in the presence of constant noise: A case study on cancer detection. *Journal of Statistical Mechanics: Theory and Experiment*, 2020(2), 023406.

Ahmad, S., Alam, M., Rahmat, M., Mubarik, M., & Hyder, S. (2022). Academic and administrative role of artificial intelligence in education. *Sustainability*, 14(3), 1101. https://doi.org/10.3390/su14031101.

Aiello, L.C. (2016). Ada Lovelace: The making of a computer scientist. *AI and Society*, 31(4), 505–513.

Al-Worafi, Y., Hermansyah, A., Goh, K., & Ming L. (2023). Artificial intelligence use in university: Should we ban chatGPT? *Preprints*, 2023, 2023020400. doi.org/10.20944/preprints202302.0400.v1.

Awad, E., Dsouza, S., Kim, R., Schulz, J., Henrich, J., Shariff, A., ... & Rahwan, I. (2018). The moral machine experiment. *Nature*, 563(7729), 59–64.

Bengio, Y., Simard, P., & Frasconi, P. (1994). Learning long-term dependencies with gradient descent is difficult. *IEEE Transactions on Neural Networks*, 5(2), 157–166. doi:10.1109/72.279181.

Benos, L., Tagarakis, A.C., Dolias, G., Berruto, R., Bochtis, D., Gemtos, T.A., & Fountas, S. (2021). Machine learning in agriculture: A comprehensive updated review. *Sensors*, 21(11), 3758.

Berrondo-Otermin, J.A., & Sarasa-Cabezuelo, A. (2023). Fighting against fake news: A Spanish initiative. *Information Polity*, 28(1), 43–55.

Bigman, Y.E. (2020). Teaching a machine to take the wheel: Ethical issues in the design of autonomous vehicles. *Science and Engineering Ethics*, 26(2), 511–533.

Boole, G. (1854/1999). *An Investigation of the Laws of Thought: On Which are Founded the Mathematical Theories of Logic and Probabilities*. Open Court Publishing.

Brown, T.B., Mann, B., Ryder, N., Subbiah, M., Kaplan, J., Dhariwal, P., ... & Amodei, D. (2020). Language models are few-shot learners. *arXiv preprint* arXiv:2005.14165.

Bullock, J.B. (2019). The future of knowledge work in the age of AI: How machine learning and artificial intelligence will transform professions. *Futures*, 112, 102422.

Canadian Institute for Advanced Research. (2022). *Pan-Canadian Artificial Intelligence Strategy*. Retrieved from https://cifar.ca/ai/pan-canadian-artificial-intelligence-strategy/

Cazzaniga, F., Coda Moscarola, F., Colombino, U., Leombruni, R., Nascia, L., & Ughetto, E. (2024). *Gen-AI: Artificial Intelligence and the Future of Work*. International Monetary Fund.

Chaudhry, A.U., & Kazim, S.A. (2021). Predicting student performance using machine learning techniques. *Computers & Education*, 168, 104207.

Chen, J. (2023). Algorithmic recruitment: Examining the impacts on fairness and efficiency in hiring practices. *Journal of Human Resources*, 58(2), 455–480.

Chernyshev, M., Zeadally, S., & Baig, Z.A. (2018). Healthcare data breaches: Implications for digital forensic readiness. *Journal of Medical Systems*, 43(1). doi.org/10.1007/s10916-018-1123-2.

Church, A. (1932). A set of postulates for the foundation of logic. *Annals of Mathematics*, 33(2), 346–366.

Church, A. (1936). An unsolvable problem of elementary number theory. *American Journal of Mathematics*, 58(2), 345–363.

Cioffi, R., Travaglioni, M., Piscitelli, G., Petrillo, A., & Felice, F.D. (2020). Artificial intelligence in industry 4.0: Applications and perspectives. *Sustainability*, 12(11), 4920.

Comunale, M., & Manera, M. (2024). *The Economic Impacts and the Regulation of AI: A Review of the Academic Literature and Policy Actions*. International Monetary Fund.

Cui, R., Xu, J., & Razzaq, A. (2022). Big data analytics in healthcare: A systematic literature review. *Technological Forecasting and Social Change*, 170, 120894.

Cybenko, G. (1989). Approximation by superpositions of a sigmoidal function. *Mathematics of Control, Signals, and Systems*, 2(4), 303–314.

De Stefano, V. (2018). *Negotiating the Algorithm: Automation, Artificial Intelligence, and Labour Protection*. International Labour Organization.

Dempsey, G., McBride, S., & Bryson, J.J. (2021). A global perspective on AI regulation: Challenges and recommendations. *AI and Society*, 36(2), 145–161.

Dilek, S., Çakir, H., & Aydin, M. (2015). Applications of artificial intelligence techniques to combating cyber crimes: A review. *Computers & Electrical Engineering*, 40(1), 20–34.

Domingues, R., Pereira, F., Moreira, F., & Prottes, D. (2020). The impact of automation on workplace safety: A systematic review. *Safety Science*, 128, 104742.

Du, Y., Lin, J., Lv, Z., Li, D., & Ding, X. (2022). AI-driven criminal investigation: A review of techniques and applications. *Forensic Science International: Digital Investigation*, 39, 301306.

Dutt, T., Singh, M., Gaur, G., & Sahu, V. (2022). Artificial Intelligence as an intelligent tool for the future animal production and better management: A review. *Agricultural Reviews*. doi.org/10.18805/ag.r-2297.

Dwivedi, Y.K., Hughes, D.L., Ismagilova, E., Aarts, G., Coombs, C., Crick, T., ... & Raman, R. (2021). Artificial Intelligence (AI): Multidisciplinary perspectives on emerging challenges, opportunities, and agenda for research, practice, and policy. *International Journal of Information Management*, 57, 101994.

Dziatkovskii, M. (2023). AI in e-learning: Current trends and future perspectives. *International Journal of Educational Technology in Higher Education*, 20(1), 12.

Elbehri, A., & Chestnov, O. (2021). *Digital Agriculture in Action: AI for Agriculture*. Food and Agriculture Organization.

Encyclopædia Britannica Inc. (2024). *Turing, Alan*. Retrieved from https://www.britannica.com/biography/Alan-Turing.

Erdélyi, O.J., & Goldsmith, J. (2018). Regulating artificial intelligence. *Proceedings of the 2018 AAAI/ACM Conference on AI, Ethics, and Society*. doi.org/10.1145/3278721.3278731.

European Commission. (2021). *Proposal for a Regulation of the European Parliament and of the Council Laying Down Harmonised Rules on Artificial Intelligence (Artificial Intelligence Act)*. Retrieved from https://eur-lex.europa.eu/legal-content/EN/TXT/?uri=CELEX%3A52021PC0206.

European Commission. (2024). *AI Liability Directive*. Retrieved from https://eur-lex.europa.eu/legal-content/EN/TXT/?uri=CELEX%3A52024PC0123.

Ezanno, P. (2021). Predictive epidemiology and precision veterinary medicine: New horizons for animal health. *Veterinary Research*, 52(1), 81.

Fang, Y., Huang, X., Liu, Y., Fu, Z., Zhong, X., & Li, X. (2023). AI in urban waste management: Trends and applications. *Environmental Science and Pollution Research*, 30(3), 4235–4247.

Fehr, R., Lee, Y., Stefan, S., Detert, J.R., Krajewski, S., Zhang, H., & Huang, J. (2022). AI-enabled decision-making: Understanding and mitigating ethical risks. *Journal of Business Ethics*, 179(3), 533–555.

Feng, J., & Ye, Q. (2021). AI and e-commerce: Applications, challenges, and future trends. *Electronic Commerce Research and Applications*, 44, 101010.

Ferrández-Pastor, F.J., Alcañiz-Lucas, M., García-Chamizo, J.M., & Platero-Horcajadas, D. (2019). AI-driven automation in agriculture: A review. *Computers and Electronics in Agriculture*, 163, 104834.

Fukushima, K. (1969). Visual feature extraction by a multilayered network of analog threshold elements. *IEEE Transactions on Systems Science and Cybernetics*, 5(4), 322–333. doi:10.1109/TSSC.1969.300225.

Fukushima, K. (1980). Neocognitron: A self-organizing neural network model for a mechanism of pattern recognition unaffected by shift in position. *Biological Cybernetics*, 36(4), 193–202. doi:10.1007/BF00344251.

Gavai, A.K. (2022). AI applications in food security: Challenges and opportunities. *Computers and Electronics in Agriculture*, 196, 106864.

Gmyrek, P., Berg, J., & Bescond, D. (2023). *Generative AI and Jobs: Policies to Manage the Transition*. International Labour Organization.

Goodfellow, Ian J., Jean Pouget-Abadie, Mehdi Mirza, Bing Xu, David Warde-Farley, Sherjil Ozair, Aaron C. Courville, & Yoshua Bengio. (2014). Generative Adversarial Nets. Neural Information Processing Systems.

Gottesman, O., Johansson, F., Komorowski, M., Faisal, A., Sontag, D., Doshi-Velez, F., & Celi, L.A. (2019). Guidelines for reinforcement learning in healthcare. *Nature Medicine*, 25(1), 16–18.

Gururaj, N. (2021). AI in e-commerce: Enhancing customer experience through intelligent systems. *Journal of Retailing and Consumer Services*, 58, 102338.

Hamamoto, R. (2022). AI for public health surveillance: Opportunities and challenges. *Journal of Biomedical Informatics*, 128, 103995.

Hanlon, G., & Jones, K. (2023). Data privacy in the digital age: Examining the implications for individuals and organizations. *Information Systems Journal*, 33(1), 123–145.

Hochreiter, S. (1991). *Untersuchungen zu dynamischen neuronalen Netzen*. [Unpublished doctoral dissertation]. Technische Universität München.

Hochreiter, S., & Schmidhuber, J. (1997). Long short-term memory. *Neural Computation*, 9(8), 1735–1780.

Hochreiter, S., Bengio, Y., Frasconi, P., & Schmidhuber, J. (2001). Gradient flow in recurrent nets: The difficulty of learning long-term dependencies. *In*: S.C. Kremer & J.F. Kolen (Eds.), *A Field Guide to Dynamical Recurrent Neural Networks* (pp. 237–243). IEEE Press.

Homma, T., Les A., & Robert M. (1988). An artificial neural network for spatio-temporal bipolar patters: Application to phoneme classification. *Advances in Neural Information Processing Systems*, 1, 31–40.

Imran, M.A., Zaman, S.I., Imtiaz, M.T., Fayaz, M., & Gwak, J. (2021). AI in healthcare: Current applications and future directions. *IEEE Access*, 9, 142014–142035.

Inter-Agency Working Group on Artificial Intelligence. (2024). *United Nations System White Paper on AI Governance*. United Nations.

Japanese Government. (2019). *AI Utilization Principles.* Ministry of Internal Affairs and Communications. Retrieved from https://www.soumu.go.jp/main_sosiki/joho_tsusin/eng/AI_utilization.html.

Japanese Government. (2023). *AI R&D and Implementation Strategy.* Ministry of Economy, Trade, and Industry. Retrieved from https://www.meti.go.jp/english/policy/economy/innovation_policy/ai_policy.html.

Jiang, Z. (2022). Prediction and management of regional economic scale based on machine learning model. *Wireless Communications and Mobile Computing*, 2022, 1–12. https://doi.org/10.1155/2022/2083099.

Jobin, A., & Ienca, M. (2019). The global landscape of AI ethics guidelines. *Nature Machine Intelligence*, 1(9), 389–399.

Kaminskyi, D., Nehrey, M., & Zomchak, V. (2021). AI in finance: Applications and challenges. *Financial Innovation*, 7(1), 64.

Kelly, C.J., Karthikesalingam, A., Suleyman, M., Corrado, G., & King, D. (2019). Key challenges for delivering clinical impact with artificial intelligence. *BMC Medicine*, 17(1), 195.

Kim, J., & Lee, S. (2022). Are two heads better than one?: The effect of student-AI collaboration on students' learning task performance. *Techtrends*, 67(2), 365–375. doi.org/10.1007/s11528-022-00788-9.

King, S., Aggarwal, N., Taddeo, M., & Floridi, L. (2018). Artificial intelligence crime: An interdisciplinary analysis of foreseeable threats and solutions. *Science and Engineering Ethics*, 24(3), 537–565.

Lee, D., Mohr, D.C., Street, R.L., & Nadkarni, P.M. (2019). Intelligent systems for clinical trials recruitment: A review. *Journal of Biomedical Informatics*, 90, 103094.

Lee, H., Chen, T.P., Chang, K.T., & Chen, H.P. (2018). Data privacy concerns and behavioral intentions: The case of mobile apps. *International Journal of Mobile Communications*, 16(6), 633–651.

Li, Y., & Wu, M. (2015). The LSTM neural network model and its application to sequence labeling. *Journal of Computer Applications*, 35(1), 15–19.

Lim, S., Edelenbos, J., & Gianoli, A. (2019). Sustainability transitions in the making: An analysis of smart city projects in the Netherlands. *Technological Forecasting and Social Change*, 142, 218–234.

Linnainmaa, S. (1976). Taylor expansion of the accumulated rounding error. *BIT Numerical Mathematics*, 16(2), 146–160.

Liu, S. (2020). Governing artificial intelligence in China: Policy and regulation. *Regulation & Governance*, 14(4), 672–687.

Liu, Z., Cheng, P., Zhang, H., Shen, W., & Mark, J.W. (2014). Autonomous vehicle-to-vehicle collision avoidance system. *IEEE Transactions on Vehicular Technology*, 63(9), 4584–4596.

Liu, X., Hentenryck, P.V., & Zhao, J. (2019). Urban traffic management with connected and autonomous vehicles: A review. *Transportation Research Part C: Emerging Technologies*, 100, 161–177.

Lopez, D.R., Tucker, A.N., Salameh, M.A., & Tucker, C.A. (2018). Personalized medicine: Genetic and epigenetic interactions in complex diseases. *Nature Reviews Genetics*, 19(3), 181–192.

López Rico-Olarte, L., Blobe, G.C., & Hullin, M.B. (2022). AI and economic development: Implications for low- and middle-income countries. *Journal of Economic Development*, 47(1), 21–34.

Ma, X., Zhang, H., Li, Q., Wang, Z., Zhao, H., & Yang, J. (2022). AI-enhanced electrical automation controls for industrial applications. *IEEE Transactions on Industrial Informatics*, 18(1), 467–478.

Marvin, H.J., Kleter, G.A., Wauters, E.A., Peeters, F.P., Dekkers, S., Beekman, A.J., ... & Houben, G.F. (2022). AI applications in food and agricultural systems: Benefits and challenges. *Trends in Food Science & Technology*, 123, 186–196.

Masood, R., Khan, M.S., Tariq, M.I., Bhatti, S., Malik, M.S., Rehman, M.H., ... & Meraj, M. (2022). AI-enabled deepfake detection: A comprehensive review. *IEEE Access*, 10, 43534–43556.

Mendoza, A.B., Hutajulu, T., & Lubis, H. (2022). AI in e-learning: Implementation and evaluation of a recommendation system. *Journal of E-Learning and Knowledge Society*, 18(2), 85–97.

Metcalf, J. (2022). Autonomous systems in military applications: Ethical implications and accountability. *Journal of Military Ethics*, 21(2), 109–129.

Metz, R. (2021). The evolution of AI: From Turing to today. *AI Magazine*, 42(1), 56–68.

Mijwil, M.M., Aggarwal, A., Mutar, M.A., Mansour, R.F., & Singh, G. (2022). AI and immersive technologies in education: Enhancing learning through augmented reality and virtual reality. *Computers and Education*, 179, 104426.

Milojević, S., & Redžepagić, S. (2021). AI applications in financial risk management: Challenges and opportunities. *Journal of Risk and Financial Management*, 14(2), 65.

Minsky, M., & Papert, S. (2017). *Perceptrons: An Introduction to Computational Geometry* (Expanded Edition). MIT Press.

Narayanan, R.R., Durga, N., & Nagalakshmi, S. (2022). Impact of artificial intelligence (AI) on drug discovery and product development. *Indian Journal of Pharmaceutical Education and Research*, 56(3s), s387–s397. https://doi.org/10.5530/ijper.56.3s.146 .

National People's Congress. (2023). *Personal Information Protection Law*. Retrieved from http://www.npc.gov.cn.

Nawaz, F., Hassan, M., Rehman, A.U., Shah, M.A., Wahid, A., Lee, S., & Khan, M.A. (2023). Privacy preservation techniques in healthcare: A comprehensive survey. *IEEE Access*, 11, 32434–32456.

Neo, H., Leow, Y., Chan, C.H., Su, Y., Chua, Y.T., & Ong, S.K. (2022). AI-driven educational games: Enhancing learning and engagement. *Educational Technology Research and Development*, 70(2), 319–340.

NIST. (2022). *AI Risk Management Framework*. National Institute of Standards and Technology. Retrieved from https://www.nist.gov/ai-risk-management-framework.

Office of the Privacy Commissioner of Canada. (2023). *Guidance on AI and Privacy*. Retrieved from https://www.priv.gc.ca/en/privacy-topics/technology-and-privacy/artificial-intelligence/

Organisation for Economic Co-operation and Development (OECD). (2023). *Recommendation of the Council on Artificial Intelligence*. Retrieved from https://legalinstruments.oecd.org/en/instruments/OECD-LEGAL-0449.

Ozgur, G., & Akkoç, H. (2021). Predicting inflation in emerging economies using machine learning techniques: Evidence from Turkey. *Economic Modelling*, 102, 105596.

Paliwal, P., Patel, J., Kandale, S., & Anute, R. (2021). AI in banking: Transforming customer experience and operational efficiency. *Journal of Banking and Financial Technology*, 5(3), 163–179.

Pappada, S. (2021). *Machine Learning in Medicine: It has arrived, let's embrace it*. doi.org/10.22541/au.163250699.94601432/v1.

Pastor, J.M., Alcañiz-Lucas, M., García-Chamizo, J.M., & Platero-Horcajadas, D. (2019). AI in smart agriculture: Trends and future directions. *Computers and Electronics in Agriculture*, 162, 104835.

Paul, A.V., Wang, S., Hu, X., Su, C., Liu, Y., & Chen, S. (2021). AI applications in drug discovery: A review. *Current Topics in Medicinal Chemistry*, 21(10), 883–902.

Petit, N. (2017). Law and regulation of artificial intelligence and robots: Conceptual framework and normative implications. *European Journal of Law and Technology*, 8(2), 81–98.

Piccarozzi, M., Aquilani, B., & Gatti, C. (2018). Industry 4.0 in management studies: A systematic literature review. *Sustainability*, 10(10), 3821.

Popenici, S.A.D., & Kerr, S. (2017). Exploring the impact of artificial intelligence on teaching and learning in higher education. *Research and Practice in Technology Enhanced Learning*, 12(1), 22.

Priya, R., Paramasivam, S., Kanagaraj, G., & Paramasivan, S. (2023). AI-driven traffic management for emergency vehicle priority. *Journal of Intelligent Transportation Systems*, 27(1), 23–34.

Randall, B. (1973). *The Origins of Digital Computers*. Berlin: Springer-Verlag.

Rao, S.N., P, K., Krishna, K.M., & Kumar, A. (2021). Data privacy concerns in the digital age: A review. *Journal of Information Security and Applications*, 58, 102887.

Rensfeldt, A.B., & Rahm, L. (2022). AI in education: Automated decision-making and administrative work. *Education and Information Technologies*, 27(3), 3859–3874.

Rios-Campo, J.P., Ortega, A., Gómez-Restrepo, C., Osorio, D.I., Bernal-Torres, C.A., & Patiño, D. (2023). AI in education: Challenges and opportunities in Colombian higher education. *IEEE Access*, 11, 38134–38148.

Rocha, H.A., Carvalho, C.V., & Aguiar, R.T. (2021). AI in healthcare: Opportunities and challenges in the Brazilian context. *Journal of Healthcare Informatics Research*, 5(2), 233–256.

Rumelhart, D.E., Hinton, G.E., & Williams, R.J. (1986). Learning representations by back-propagating errors. *Nature*, 323(6088), 533–536.

Sahu, A.K., & Kumar, S. (2022). AI in healthcare: Opportunities, challenges, and future directions. *Journal of Biomedical Informatics*, 127, 103978.

Salehin, M.M. (2024). Automated Machine Learning: Techniques, tools, and challenges. *Artificial Intelligence Review*, 53(1), 1–29.

Santos, J. (2023). AI in combating disinformation: Techniques and challenges. *Information Processing & Management*, 60(2), 102967.

Savoia, R. (2023). Data-driven decision-making in public health: Leveraging AI for improved outcomes. *Health Policy and Technology*, 12(1), 100431.

Schiff, D. (2020). AI in education: Promises, challenges, and future directions. *AI & Society*, 35(3), 605–615.

Sepp, H., & Jürgen, S. (1997). Long short-term memory. *Neural Computation*, 9(8), 1735–1780.

Shi, Y. (2022). AI in logistics: Opportunities and challenges. *Journal of Supply Chain Management*, 58(2), 123–135.

Singh, A.K., Singh, P., Vege, S., & Singh, H. (2022). AI in cybersecurity: Enhancing threat detection and response. *Journal of Cyber Security and Mobility*, 11(2), 177–196.

State Council of China. (2021). *New Generation Artificial Intelligence Development Plan*. Retrieved from http://english.www.gov.cn/policies/latest_releases/202107/20/content_WS5f15b5a4c6d03f32.html.

Stach, P. (2023). Data is the new oil: Understanding the role of data in the digital economy. *Journal of Information Systems*, 37(1), 45–58.

Su, Y., & Yang, Y. (2023). Chatbots in education: Enhancing learning experiences through AI. *Computers and Education*, 179, 104425.

Sumangala, K.R., & Kini, A. (2022). AI in agriculture: Innovations and challenges. *Journal of Agricultural Science and Technology*, 24(1), 123–134.

Tealab, Ahmed. (2018). Time series forecasting using artificial neural networks methodologies: A systematic review. *Future Computing and Informatics Journal*, 3(2), 334–340. doi:10.1016/j.fcij.2018.10.003. ISSN 2314-7288.

Treasury Board of Canada Secretariat. (2021). *Directive on Automated Decision-Making*. Retrieved from https://www.tbs-sct.gc.ca/pol/doc-eng.aspx?id=32592.

Trandabăț, D., & Gifu, D. (2023). Generating synthetic text corpora for AI training: A case study on fake news. *Journal of Artificial Intelligence Research*, 68, 231–255.

Turing, A.M. (1950). Computing machinery and intelligence. *Mind*, 59(236), 433–460.

Um, J., Wu, F., & Zhang, A. (2014). Housing value forecasting based on machine learning methods. *Abstract and Applied Analysis*, 2014, 1–7. doi.org/10.1155/2014/648047.

United Nations Children's Fund (UNICEF). (2021). *Policy Guidance on AI for Children*. Retrieved from https://www.unicef.org/globalinsight/reports/policy-guidance-ai-children.

United Nations Educational, Scientific, and Cultural Organization (UNESCO). (2022). *Recommendations on the Ethics of AI*. Retrieved from https://unesdoc.unesco.org/ark:/48223/pf0000377897.

United Nations Educational, Scientific, and Cultural Organization (UNESCO). (2022). *Guidance for Generative AI in Education and Research*. Retrieved from https://unesdoc.unesco.org/ark:/48223/pf0000377897.

United Nations General Assembly. (2024). *Harnessing the Opportunities of Safe, Secure, and Trustworthy Artificial Intelligence Systems for Sustainable Development*. Retrieved from https://undocs.org/A/78/L.49.

Vaswani, A., Shazeer, N., Parmar, N., Uszkoreit, J., Jones, L., Gomez, A.N., ... & Polosukhin, I. (2017). Attention is all you need. *In*: *Proceedings of the 31st Conference on Neural Information Processing Systems* (*NIPS 2017*), 5998–6008.

Velasco, F.J. (2022). AI-enhanced malware and ransomware: Threats and defenses. *Journal of Information Security*, 13(2), 65–78.

Wahl, B., Cossy-Gantner, A., Germann, S., & Schwalbe, N.R. (2018). Artificial intelligence (AI) and global health: Opportunities and challenges. *Global Health*, 14(1), 111.

Wang, Y., Sun, S., & Zhou, L. (2023). Enhancing e-learning through AI: Applications and future directions. *Journal of Educational Computing Research*, 59(3), 538–559.

Wasilow, C., & Thorpe, T. (2019). Autonomous systems in warfare: Addressing ethical and humanitarian challenges. *International Review of the Red Cross*, 101(910), 329–351.

Werbos, P.J. (1982). Applications of advances in nonlinear sensitivity analysis. *In*: *System Modeling and Optimization* (pp. 762–770). Springer.

White House Office of Science and Technology Policy. (2023). *Blueprint for an AI Bill of Rights*. Retrieved from https://www.whitehouse.gov/ostp/ai-bill-of-rights/

World Health Organization. (2021). *Ethics and Governance of Artificial Intelligence for Health*. Retrieved from https://www.who.int/publications/i/item/9789240029200.

World Health Organization. (2023). *Regulatory Considerations on Artificial Intelligence for Health*. Retrieved from https://www.who.int/publications/i/item/9789240029200.

World Health Organization. (2024). *Ethics and Governance of Artificial Intelligence for Health: Guidance on Large Multi-Modal Models*. Retrieved from https://www.who.int/publications/i/item/9789240029200.

Wu, G., & Ge, Y. (2019). Intelligent logistics management: AI applications and challenges. *Journal of Logistics Management*, 11(1), 25–36.

Xia, F., Yang, L., Huang, Y., & Huang, X. (2019). AI in e-commerce: Enhancing customer experience through intelligent systems. *Journal of Retailing and Consumer Services*, 50, 46–56.

Yedla, A., Kakhki, F., & Jannesari, A. (2020). Predictive modeling for occupational safety outcomes and days away from work analysis in mining operations. *International Journal of Environmental Research and Public Health*, 17(19), 7054. doi.org/10.3390/ijerph17197054.

Yiğitcanlar, T., Kankanamge, N., Vella, K., Sabatini-Marques, J., & Dawes, L. (2020). Smart cities and disaster management: An overview of AI applications and challenges. *International Journal of Disaster Risk Reduction*, 50, 101789.

Zakon, A. (2019). Social media addiction: The role of AI in capturing user attention. *Journal of Social Media in Society*, 8(2), 230–246.

Zawacki-Richter, O., Marín, V.I., Bond, M., & Gouverneur, F. (2019). Systematic review of research on artificial intelligence applications in higher education: Lessons and recommendations for the field. *International Journal of Educational Technology in Higher Education*, 16(1), 39.

Zhang, L., Wang, Y., & Zhu, H. (2022). Theory and experiment of cooperative control at multi-intersections in intelligent connected vehicle environment: Review and perspectives. *Sustainability*, 14(3), 1542. doi.org/10.3390/su14031542.

Zheng, Q., & Badarch, D. (2022). AI in education: Administrative applications and ethical considerations. *International Journal of Educational Technology in Higher Education*, 19(1), 45.

Zhou, X., Zhang, Y., Luo, Z., Parker, R.B., & Choudhury, N.R. (2023). Creating synthetic text corpora to combat fake news: Methodologies and challenges. *Journal of Information Science*, 49(2), 153–170.

Zhu, H., Shi, Y., & Liu, Y. (2022). AI in industry: Enhancing productivity and operational efficiency. *Journal of Manufacturing Systems*, 62, 212–223.

Zhu, J., & Zheng, W. (2021). AI in the judiciary: Enhancing efficiency and accountability. *Journal of Legal Studies*, 50(1), 123–146.

2 | The Brazilian Bioethics of Protection: Exploring the Old and the New

Fermin Roland Schramm

Bioethics, understood as the study of the morality of human acts that have real or potential irreversible consequences on any living system, refers to both emerging and persistent health issues of individuals and human populations in the context of their interrelations with each other and their natural, cultural, and social environments. This essay has two main objectives: (1) to describe the emergence of bioethics in Brazil, which we qualify as both young and late due to its context of emergence; (2) to provide a brief critical review of its production to define the conditions for its integration into the context of global bioethics, highlighting the possibility of it being a complex bioethics capable of addressing both persistent and emerging health issues, based on decades of research and institutional activity.

Introduction

In this text, I intend to address two subjects: (1) to characterize bioethical research in Brazil from the health context in which it manifests, defining its traits pertinent to evaluating potential moral conflicts involved; (2) to outline what could be its contribution to a possible global bioethics, consisting of the ability to deal with a toolbox that has a history but must reinterpret it in light of the characteristics of Brazilian problems, such as health issues; that is, reformulating the bioethical problem in complex terms (to be able to address its concrete problems), and pragmatic terms (to optimize the relationship between means and ends of public policies in terms of efficiency and effectiveness, particularly in the health field).

I believe that if there is any original contribution from Brazilian bioethicists to the field of bioethical reflections and solutions as a whole (which we could

Email: rolandschram@yahoo.com.br

call "global bioethics"), it will only be possible through competence in wisely addressing both "persistent problems" (such as public health issues and lack of care and equity), and "emerging problems" (such as those implicated by biotechnoscientific practices like genetic engineering and robotics) embedded in their moral reality (Garrafa, Costa, & Oselka, 2000).

In other words, "bioethics in Brazil" will only move from a possible critical condition of minority to a desirable adult status of "Brazilian bioethics" with a sufficiently unique identity to be inscribed in its cultural and political tradition. At least, it will be able to adapt the 'imported' models, which some may see as indispensable (as they would be part of the toolbox still used, for example, in biomedicine) but others may see as insufficient (as has been seen in recent decades in the case of principlism), taking care not to practice any form of reductionism (such as the universal use of the biomedical model in the entire health field) nor forms of Confucianism (resulting from the lack of clarity in constructed ideas and produced discourses), as the 'global' scope in which bioethical problems are inscribed arises at different pertinent levels of analysis and coherent proposals for solutions.

In particular, approaches to "persistent problems" and "emerging problems" are both relevant to solving human health and well-being issues in their concrete contexts, as a negligent approach to "emerging problems" (e.g., genetic predispositions to preventable diseases through predictive tests) can make them 'persistent' in the future. In this case, we would have moral responsibility for our negligence, especially towards future generations, as it is reasonable to state that we are equally responsible for actions and omissions (Glover, 1977); that is, we are morally responsible both for the harm we cause and for the harm we could have avoided. In fact, as the English bioethicist John Harris stated at the end of the last century,

> If we are to have any chance of maintaining or achieving control over our destiny, we must try to anticipate what might happen [as future dangers tend to become] infallibly actual and real, and when they do, they may be more difficult to control [which] would make us responsible for causing immense avoidable suffering (Harris, 1998).

On the other hand, the scarcity of resources to satisfy different spheres of legitimate goods (such as health and well-being) is not always truly available in a society, making it impossible for any health system in the world to easily meet all the health needs of its population. Otherwise, it would probably (and inevitably) go bankrupt; this implies setting some limits on the increasing demands for health services, especially those that can be considered in principle consensually as unreasonable compared to others also considered consensually more important and even essential (Callahan, 1987). But establishing which needs are 'basic' and which are 'accessory', that is, which would have lexical priority over others, cannot evidently be established solely by the confrontation of rational and sufficiently justified arguments, as this presupposes that all interest groups in a society share the same type of rational analysis or, as discourse ethics theorists (especially Apel and Habermas) argue, some *a priori* (transcendental or not) condition that guarantees the necessary interlocution to consensually construct a solution considered ethically (and politically) correct. In practice, this is far from being more than a societal project, especially if we consider the factual existence of a plurality of 'reasons' that may often have no common denominator, thus requiring forms of argumentative and political mediation that may not always and necessarily be rational.

Here, we highlight the case of the bioethics of protection (BP) as an example of Brazilian bioethics. We identify BP as:

> "an applied ethics concerning human practices that can have significant irreversible effects on living beings and particularly on individual and population humans. These practices are considered in their ecological, biotechnoscientific, and sociocultural contexts, considering the conflicts of interests and values that may emerge in such practices. BP intends to account for and, as far as possible, resolve such conflicts."

In more detail, BP deals with:

(a) Describing and understanding them in the most rational and impartial way possible.
(b) Resolving them by proposing tools that can be considered by any rational and reasonable moral agent as the most adequate to proscribe behaviors considered incorrect and prescribe those considered correct.
(c) Thanks to the correct articulation between (a) and (b), providing the means capable of sufficiently protecting those involved in such conflicts, guaranteeing each life project compatible with the others.

In fact,

> "this definition adds something to the usual definitions of bioethics. It not only tries to analyze conflicts and prescribe what should be done but also aims to rescue what is probably the primary meaning of the Greek word *ethos*, which precisely has the sense of 'shelter', 'refuge', and 'protection'; in short, protection" (Schramm & Kottow, 2001).

In particular, the applied ethics specialist adopting the bioethics of protection model must know how to ask if there is a pertinent specific object of public health. This leads to identifying a specific bioethical problem, such as the moral problem addressed by the bioethics of protection. This model deals with the moral implications of public health policies aimed mainly at prevention, which seeks to avoid health damage. It also focuses on promoting population health guaranteed by the State. The means to achieve this is the institutional responsibility to protect the community. This community is understood as the specific object of public health, according to the diffuse slogans of "disease prevention" and "health promotion".

Faced with this possible object, there is the position that considers the inappropriateness of the Principlist model to the moral problem of the health reality. It should be replaced by guiding principles such as solidarity and responsibility. This includes responsibility towards the totality of being, defended by Hans Jonas, or the radical responsibility, defended by Emmanuel Lévinas.

However, both authors do not delve into the field of their applicability to public health conflicts. In such conflicts, we should clearly identify the objectives, means, and actors of responsibility, which is not clear in Jonas. Additionally, there must be a reasonable conception of responsibility, open to all regardless of belonging to a specific moral community, which is also difficult to identify in Lévinas (Schramm & Kottow, 2001).

In particular, the concept of 'protection' (which is, incidentally, the first meaning of the Greek word *ethos*) (Chantraine, 1968) seems to be the one that best adapts to the type of responsibility required by the moral conflicts of public health, as it aims to integrate moral responsibility and efficiency in respecting both the plurality of needs and values of current multicultural and multi-confessional societies and a reasonable way of providing health goods to those who need them.

Thus, a bioethics guided by the principle of protection recovers the original sense of ethics, rooted in the tradition of social medicine and the role of the modern State. Starting from the idea that a State is only legitimate if it offers protection to its citizens, who may accept it or not without being forced to do so, which makes all the difference between a protective policy and a paternalistic policy.

Characteristics and Context of Bioethics in Brazil

The construction of bioethics in Brazil began approximately three decades ago, and its existence as a set of academic and institutional activities can be considered capable of treating Brazilian bioethics as simultaneously 'young' and 'late'. This is because it is concerned with understanding and resolving old and new challenges in the field of human health and biomedical research, dealing simultaneously with both 'persistent' and 'emerging' problems of individual and collective health.

It is 'young' because it emerged in the early 1990s, almost a century after the term Bio-ethik was coined in a 1927 essay by the German theologian Fritz Jahr to name a new ethical imperative regarding behavior towards all living beings (not only humans), asking the question:

> "The 5th commandment warns: '*Thou shalt not kill.*' The term kill always means killing something alive. Living beings are humans, animals, and plants. Since the 5th commandment does not prohibit killing exclusively humans, should it not be applied to animals and plants analogously?" And

> answering, "Respect every living being as a matter of principle [or as an end in itself] and treat it if possible as such" (Jahr, 1927).

And it developed more than 40 years after the Nuremberg Code (1946), which established the first international ethical guidelines for research with human beings, and 20 years after the official (nominal) birth of bioethics in the field of applied ethics.

In fact, bioethics was officially born in the United States when the American oncologist Van Rensselaer Potter recreated the neologism bioethics to indicate a new field of research and action in traditional biomedical ethics based on the traditional Hippocratic principles. For this author, this new field would have an interdisciplinary character with a marked ecological content (thus recovering the discussion launched by the theologian Jahr) and should be concerned with the survival of the human species and the responsibility of scientists towards the present quality of *Lebenswelt*[1] (Potter, 1970).

Two decades later, anticipating many current debates and concerns in environmental bioethics and biosafety, Potter would rename his initial conception as "global bioethics" (Potter, 1988). He maintained the contents of the original conception, namely: (a) the promotion of human dignity (to deter possible abuses of technoscience); (b) voluntary fertility limitation (to avoid negative pressures on global quality of life due to overpopulation); and (c) respect for the natural environment (to stop the degradation of conditions necessary for the existence of life itself).

The author's intention was to distinguish his broad conception of bioethics. This conception was simultaneously bio-logical, interdisciplinary, and anthropo-cosmic. He wanted to differentiate it from competing conceptions, most of which were considered limited and insufficient. He distinguished it, in particular, from the dominant conception in the biomedical field during the 1970s and 1980s known as Principlism (Beauchamp & Childress, 1989).

Brazilian bioethics emerged in this 'globalized' and 'dynamic' context (Anjos, 1997). At this time, the main issues on the current bioethical discussion agenda (abortion, euthanasia, organ transplantation, medical iatrogeny, resource allocation, etc.) were already practically established. However, Brazilian bioethics brought its own problems linked to public health and resulting from injustice and social exclusion to this discussion agenda.

In particular, there is the bioethical problem of 'misery', which is perhaps the most fundamental problem of ethics from a social perspective. It concerns the necessary conditions for the quality of life of individuals and populations. It

---

1 Editor's note: Lebenswelt is a term from German philosophy, particularly in phenomenology, meaning 'lifeworld'. It refers to the pre-theoretical, everyday world of lived experiences and interactions, encompassing the social, cultural, and natural contexts in which people exist and perceive their surroundings directly. Introduced by Edmund Husserl, it contrasts with the objective world studied by sciences.

refers to the human condition in its essential fragility (Schramm, 1996) or, more radically, as it is, in fact, a "radical death threat not to men but to Humanity in man" (Leão, 1992).

Additionally, there are problems related to the precarious protection of health for both individuals and populations (Schramm & Kottow, 2001). These issues are implicated by conflicts related to the allocation of scarce resources. This occurs in the context of a 'tyrannical' reduction of all spheres of distributive justice in a pluralistic and complex society (as is, in principle, the Brazilian one) to the economic sphere (Walzer, 1983).

It is 'late' because, when it emerged internationally, an important paradigmatic transition in bioethics was already clearly delineated. This transition was characterized, on the one hand, by a shift from a bioethics that we could call essentially 'private'. This 'private' bioethics was marked by a gaze predominantly concerned with the ethical problems resulting from the acts 'of' and 'on' individual humans. It focused on the so-called first-generation or 'personal' human rights. On the other hand, there was a shift to a 'public' bioethics more concerned with the ethical dimension of collective health problems. This 'public' bioethics was based on the so-called social or second-generation rights (Garrafa, 1995).

Additionally, the complexification of its field of study and action was due to what we could call the "hyper health crisis". This term indicates the synergy between old and new health problems. It refers to 'persistent' health problems largely resulting from underdevelopment, such as infectious-contagious diseases and diseases from the lack of basic sanitary conditions like sanitation. It also includes 'emerging' health problems induced by scientific advances and technological incorporation in health procedures. Examples include chronic-degenerative diseases resulting from the epidemiological transition and the consequent aging of the population.

All this occurred in the context of emerging bioethical discussions related to the environment, discussed, for example, during the 1992 Rio World Conference. It also included third-generation (or "future generations") and fourth-generation (or 'environmental') rights (Schramm et al., 1992). Other issues were the impact of external debt on budget cuts and the negative consequences for the effectiveness of health policies. There was also the 'persistent' practice of institutional promiscuity between public and private and/or corporate interests, which may imply corruption and diversion of money considered essential for programs of primarily social interest.

The simultaneously 'young' and 'late' character of bioethics practiced in Brazil had some comparative disadvantages concerning bioethical research in countries with an older tradition. This was especially evident at the beginning. It implied an excessive dependence on one or another established (or 'consecrated') model of analysis.

Particularly, there was a reliance on the "Georgetown model" or the four prima facie principles (beneficence, non-maleficence, autonomy, and justice). This model is American and inspired, for example, the drafting of Resolution 196/96[2] of the Brazilian Ministry of Health, which still regulates research with human beings.

Additionally, there was some dependence, although to a lesser extent, on the American and European personalist model. This model is more concerned with highlighting the foundations of personal rights and emphasizing "the importance of personality" (Mautner, 2010).

Finally, there was also a reliance on the universalist humanism of the Enlightenment and Kantian tradition. This tradition "adopted the ideal of the full development of the individual, rejecting both religious asceticism and narrow scholasticism as well as humble piety" (Idem, p. 375).

In the specific case of the 'success' of the Principlist model, this is probably due to a relatively easier (but above all supposed) applicability of the matrix of the four principles. These principles apply to various contexts and situations of moral conflict in the biomedical field in liberal, democratic, and pluralistic societies (such as the American and, partially, Brazilian).

In the case of the personalist model, this may be due to a relative dependence of Brazilian intellectuals on European culture, particularly French culture. In this context, personalism linked to the "philosophy of probabilities" (Blackburn, 1997) developed from the 1950s with the work of the Catholic philosopher Emmanuel Mounier, who emphasized the autonomy and value of the person, opposing "dehumanizing tendencies [and] the need for necessary community and personal relationships for the flourishing of the person" (Mautner 2010, p. 513).

As for Enlightenment humanism, this is due to the long tradition of Brazilian academic philosophy dependent on the continental model, especially its French and partly German branches. In recent years, however, there has been a growing importance of analytic tradition philosophy (especially English). This, combined with German neo-Aristotelianism and American neo-pragmatism, constitutes the cultural humus from which applied ethics, particularly bioethics, emerge.

However, the fact that bioethical activities in Brazil are 'young' and 'late' has also, after an initial period of dependence on the two previous models – which lasted practically throughout the 1990s – led to a certain renewal constituted by a growing criticism of "imported models", particularly the Principlist model when applied to public health problems. In other words, the 'adolescence' of Brazilian bioethics has a comparative advantage, as it paradoxically implies greater freedom concerning existing models that can be 'cannibalized' (in the sense of

---

[2] Editor's note: The Resolution 196/96 was issued by the Brazilian National Health Council (*Conselho Nacional de Saúde*) in 1996; this resolution establishes guidelines and standards for research involving human subjects in Brazil. It aims to ensure the ethical conduct of research by setting principles such as respect for human dignity, protection of participants' rights, and the promotion of beneficence and justice. The resolution mandates the creation of Research Ethics Committees (*Comitês de Ética em Pesquisa*) to oversee and evaluate research proposals, ensuring compliance with ethical standards. It is considered a foundational document for bioethics and research ethics in Brazil.

cultural anthropophagy[3] of the São Paulo modernists) without much ceremony concerning each used tradition, which allows, in principle, the use of the full range of possibilities offered by the 'toolbox' of moral philosophy, focusing solely on argumentative consistency and the consideration of the pragmatic effects of decisions considered ethically legitimate.

But this phase, which we could call the 'adolescent' phase of Brazilian bioethics, is also reaching a moment of balance. The direction of the 'adult' phase will likely depend on this balance. The main challenge is already delineated as a complex task that consists of knowing how to distinguish without separating and unite without confusing the problems of private bioethics and public bioethics. It involves knowing how to distinguish the types and areas of relevance of the various problems to be addressed.

It is from this new competence that the credibility and legitimacy of Brazilian bioethicists on the world stage depended. This was evident during the Sixth World Congress of the International Association of Bioethics (IAB), held in Brasília in 2002.

How to Think about a Complex Brazilian Bioethics?

Due to the characteristics of Brazilian health and well-being problems outlined here, Brazil is not necessarily a "special case" (Sonderfall) on the world stage. This is unless one opts for a casuistic view, in which any case can be considered a special case. However, these characteristics make Brazil a paradigmatic case of health and well-being problems of this 'postmodern' era.

This era is associated with "the joyful acceptance of the immediate... and celebration" and "the denial of the existence of stable meanings of the correspondence between language and the world" (Blackburn 1997, p. 305). These characteristics are probably valid for 'dependent' countries in similar situations.

The case (widely reported by the media) of repeated threats to break patents of essential drugs by the Brazilian Government, as well as its policy of producing generic and similar drugs to reduce costs, is a good example of this. This approach was followed by other Third World countries and is forcing some of the large multinational pharmaceutical companies to seek agreements on prices considered more affordable to the budgets of each country (and each pocket).

---

3 Editor's note: Cultural Anthropophagy refers to the concept of "cultural cannibalism" introduced by the São Paulo modernists during the *Semana de Arte Moderna* (Modern Art Week) of 1922. It was a metaphor for the process of cultural assimilation and transformation, where Brazilian artists and intellectuals sought to 'devour' and incorporate elements of European and other foreign cultures, transforming them into something uniquely Brazilian. The idea was articulated by Oswald de Andrade in his "Manifesto *Antropófago*" (Cannibalist Manifesto) of 1928, advocating for a creative and critical assimilation of external influences to foster an authentic Brazilian culture that was both modern and rooted in local realities.

This was only possible because the attitude of the Brazilian authorities was ethically based and politically supported by the WHO itself. The WHO considers the right to health a fundamental human right.

In effect, economic reasons, although fundamental for the smooth functioning of a society (as critically analyzed by Marx before this became a neoliberal dogma), cannot subsume all other fundamental social dimensions. Doing so would create uncontrollable political and social crises precisely because they are considered 'tyrannical' by contemporary democratic and pluralistic societies (Walzer, 1983).

In these societies, health cannot be seen as a mere commodity subject to the same rules as any other consumer product. On the contrary, health must be considered a fundamental good. In principle, it has at least the same relevance as "economic health" to ensure some kind of legitimate social order, if not the most fundamental good. This is because other fundamental necessities may depend on it, as popular wisdom seems to know well.

However, this does not mean that there should not be strict control in pragmatic terms of the effectiveness of health policies. This involves the optimization of the relationship between reasonable health objectives (which define the universe of legitimate ends) and the procedures to be used to achieve these ends (which constitute the universe of the most appropriate means). This must be done in the context of the specific characteristics of each society. Otherwise, as well pointed out by Callahan, any health system would inexorably go bankrupt (Callahan, 1987).

For this reason, this problem is 'complex' and therefore deserves the wisdom to find the Archimedean point between 'economism' and *laissez-faire*. This balance makes it possible to achieve some reasonable form of distributive justice in terms of equity. It involves reducing as reasonably as possible disparities in access to existing and relevant goods from the perspective of health protection. This means prioritizing, for example, the most in need through focused health policies.

There are evidently extreme cases such as those of very poor countries (as some African countries still are) or very poor regions within a single country (as is the case of Brazil itself). However, for these cases, *mutatis mutandis*, the previous arguments may also apply, as in a supposedly 'globalized' world, the parts must be in solidarity with each other for there to be a minimally ordered whole instead of chaos.

In Brazil, on all these issues, there is a reasonable sanitary culture linked to social justice problems. Bioethical reflection in Brazil begins largely with ethics applied to public health problems. This is certainly the case, inter alia, of the Oswaldo Cruz Foundation in Rio de Janeiro, the Federal University of Rio de Janeiro, the School of Public Health of the University of São Paulo, and the Faculty of Health Sciences of the University of Brasília. This does not intend to ignore the contributions of the Forensic Medicine Faculties of some Medical Faculties and some Philosophy Faculties.

But one of the complex aspects of 'adult' bioethics may be that concerning—as already indicated above—how to relate the 'persistent, and 'emerging' health problems.

One reason is certainly the fact that it is a counterintuitive issue. It seems logical to prioritize persistent problems if one has to choose. However, many emerging problems can become persistent in the medium or long term.

Therefore, public health bioethics cannot forget the moral implications of advances in science and technology, particularly those concerning biomedicine. This includes new forms of assisted reproduction. It also includes new predictive tests created by genomic medicine (and recently proteomics) and molecular epidemiology, which studies possible interactions between genome, environment, and lifestyles. Additionally, there are new treatments arising from genetic engineering, such as the use of stem cells, organ and tissue cloning, and transgenics, among others.

In this case, the complexity of the problem requires not only wise management of available resources for health in a particular society but also a broad and informed public debate on the pros and cons of incorporating such procedures and technologies into the health system, as it is from this debate that the social legitimacy of a public policy in health also depends. The risk is evidently that ethically questionable solutions are chosen, but this is an inherent risk in any democratic system. However, it can also be assumed that in a communication society that opts to stimulate the conscious and vigilant exercise of citizenship, there may be the most appropriate means to reach reasonable and acceptable forms of consensus.

In any case, it is ultimately a bet between 'optimists' and 'pessimists', who can very well reach a compromise between the pessimism of reason and the optimism of (good) will.

Another much more serious reason concerns the fact that, in the face of emerging problems, we are almost always in a condition of uncertainty. This implies acting in risk situations, that is, having to calculate unforeseeable consequences. This is why it is extremely difficult to formulate socially acceptable public health policies based solely on "risk probabilities".

These probabilities pertinently apply to populations but not necessarily to individuals. In principle, individuals only accept to give up something by coercion or by considering themselves satisfactorily integrated into the group or society for which they would have to give up their interests and/or desires.

This evidently leads us back to the issue of social exclusion, but now seen from another perspective. Social inclusion can be seen from the perspective of risk prediction in a situation of uncertainty. It is a necessary condition for cooperation and prevention (as far as reasonably possible) of risk behaviors that may threaten others.

Final Considerations

Bioethics, broadly understood as the study of the morality of human acts that have real or potential irreversible consequences on any living system (which is always also an irreversible system), concerns both emerging and persistent health issues. These issues affect individuals and human populations in the context of their interrelations with each other and their natural, cultural, and social environments.

In this essay, I had two main objectives: (1) to describe the emergence of "bioethics in Brazil", which we qualify as both young and late due to its context of emergence; (2) to provide a brief critical review of its production to define the conditions for its integration into the international context as "Brazilian bioethics". This highlights the possibility of it being a complex bioethics capable of addressing both persistent and emerging health issues in an original way, based on its experience acquired over decades of research and institutional activity.

I believe this transition is possible. The engagement and intellectual and emotional generosity of many old and young Brazilian bioethicists, whose work indirectly inspired these modest notes, leave me hopeful. These notes aim only to indicate a possible reading path that is not a dead end, but a route with an exit, as the proposed bioethics of protection intends to be.

References

Anjos, M.F. (1997). Bioética: Abrangência e dinamismo. *O Mundo da Saúde*, 21(1), 4–12.

Beauchamp, T., & Childress, J.F. (1989). *Principles of Biomedical Ethics*. New York: Oxford University Press.

Blackburn, S. (1997). Personalismo. *In*: Jorge Zahar (Ed.), *Dicionário Oxford de filosofia* (p. 296). Rio de Janeiro. Callahan, D. (1987). *Setting Limits: Medical Goals in an Aging Society*. Washington DC: Georgetown University Press.

Callahan, D. (1998). *False Hopes: Why America's Quest for Perfect Health is a Recipe for Failure*. New York: Simon & Schuster.

Chantraine, Y.P. (1968). *Dictionnaire Ethymologique de la Langue Grecque*. Paris: Klincksieck.

Garrafa, V. (1995). *A dimensão da ética em saúde pública*. São Paulo: Ed. USP/Kellogg Foundation.

Garrafa, V., Costa, S.I.F., & Oselka, G. (2000). A bioética no século XXI. *In*: V. Garrafa & S.I.F. Costa (Eds.), *A Bioética no século XXI* (pp. 13–23). Brasília: Ed. UnB.

Glover, J. (1977). *Causing Death and Saving Lives*. Harmondsworth: Penguin Books.

Harris, J. (1998). *Clones Genes and Immortality: Ethics and the Genetic Revolution*. Oxford: Oxford University Press.

Jahr, F. (1927). Bio-ethik. Eine Umschau über die ethischen Beziehungen des Menschen zu Tiere und Pflanzen. *Kosmos*, 24(2).

Leão, E.C. (1992). A ética do desenvolvimento. *In*: M.C. Leal, P.C. Sabrosa, R.H.

Rodríguez, & P.M. Buss (Eds.), *Saúde, Ambiente e Desenvolvimento* (Vol. 1, p. 221). São Paulo: Ed. Hucitec-Abrasco.

Mautner, T. (2010). *Dicionário de filosofia*. Lisboa: Ed. 70.

Potter, V.R. (1970). Bioethics: The science of survival. *Perspectives in Biology and Medicine*, 14, 127–153.

Potter, V.R. (1988). *Global Bioethics: Building on the Leopold Legacy*. East Lansing: Michigan State University Press.

Potter, V.R. (1990). Getting to the year 3000: Can Global Bioethics overcome evolution's fatal flaw? *Perspectives in Biology and Medicine*, 34, 89–98.

Schramm, F.R., Borges, S.M.N., Valadares, J.C., Czeresnia, D., Barata, P., Valle, S., & Almeida, C.M. (1992). Ética e ecologia. Algumas reflexões comuns. *In*: M.C. Leal et al. (Eds.), *Saúde, Ambiente e Desenvolvimento* (Vol. 1, pp. 207–216). São Paulo: Ed. Hucitec-Abrasco.

Schramm, F.R. (1996). *A terceira margem da saúde: Ética natural complexidade crise e responsabilidade no saber-fazer sanitário*. Brasília: Ed. UnB.

Schramm, F.R. (1997). *Da bioética privada à bioética pública*. *In*: S. Fleury (Ed.), *Saúde e Democracia: a luta do CEBES* (pp. 227–240). São Paulo: Ed. Lemos.

Schramm, F.R. (2000). Bioética economia e saúde: Direito à assistência justiça social alocação de recursos. *Revista Brasileira de Cancerologia*, 46(1), 41–47.

Schramm, F.R., & Kottow, M. (2001). Principios bioéticos en salud pública: Limitaciones y propuestas. Cadernos de Saúde Pública, 17(4), 949–956.

Walzer, M. (1983). *Spheres of Justice: A Defense of Pluralism and Equality*. New York: Basic Books.

3 | Nations of the Global South: Pioneering the Future of Health with Artificial Intelligence and Digital and Health Sovereignty

Wellington Pinheiro dos Santos*, Maíra Araújo de Santana, Fabiano Tonaco Borges, Leandro Modolo and Giselle Machado Magalhães Moreno

Artificial Intelligence (AI) emerges as a powerful tool to transform the health sector in the Global South, driving innovation, optimizing processes, and enhancing the quality of care. However, applying AI in health presents challenges and opportunities that require critical reflection, especially in the context of digital imperialism and the pursuit of health sovereignty. This chapter explores the challenges of AI in health, such as unequal access to technology, technological dependence, brain drain, and ethical and regulatory issues. Conversely, the opportunities of AI for the Global South include strengthening health sovereignty, improving the quality of healthcare, developing innovative solutions, and promoting equity. To overcome these challenges and build a fairer health future, the authors propose a set of strategic measures: investing in AI research and development in health; creating infrastructure for health data storage and processing, training professionals specialized in AI in health; promoting South-South cooperation; and implementing robust regulatory frameworks for AI in health. Achieving digital and health sovereignty is a strategic imperative for the Global South in the digital age. By investing in research, development, and innovation in AI, together with promoting the national technology industry and building strategic partnerships, the Global South can construct a more autonomous, resilient, and just future where AI is used for the well-being of the population and sustainable development.

Introduction

The global health landscape is a complex mosaic marked by technological

*Corresponding author: wellington.santos@ufpe.br

advances, demographic changes, and persistent challenges that impact nations unevenly. For Global South countries, this scenario presents both obstacles and promising opportunities for building a healthier and more prosperous future (Islam, 2023; Kasai, 2019; Guzmán, 2022; Loewenson, 2021).

At the heart of the debate are health disparities between rich and poor countries. Developed nations enjoy greater access to quality healthcare services, robust infrastructure, and medical advances. In contrast, much of the Global South's population faces different realities. Communicable diseases like HIV/AIDS, tuberculosis, and malaria still haunt communities, while chronic non-communicable diseases such as heart disease, diabetes, and cancer gain ground, straining already fragile health systems (Mabhala, 2020; Ho & Dascalu, 2020; Scheerens et al., 2020). Lack of access to clean water, basic sanitation, and adequate nutrition worsens the situation, perpetuating a cycle of poverty and disease.

The Covid-19 pandemic highlighted global health inequalities. The virus's rapid spread and devastating impacts exposed the fragility of health systems in many Global South countries, underscoring the urgent need for greater investment in infrastructure, human resources, and preventive measures (Jain, 2021; Nguyen, 2022; Vishwanathan, 2022; Carter et al., 2020). The pandemic also accelerated the search for innovative solutions, driving the development of new technologies and promoting international collaboration in scientific research (Jain, 2021; Nguyen, 2022; Vishwanathan, 2022; Carter et al., 2020).

In this challenging context, technology emerges as a beacon of hope, offering promising tools to reduce health disparities and build more resilient and equitable health systems. Artificial Intelligence (AI), for example, can assist in data analysis for more accurate diagnoses, the development of personalized medicines and vaccines, and the expansion of access to quality healthcare through telemedicine (Secinaro et al., 2021; Amjad et al., 2023; Santus et al., 2021). Digital tools can monitor the health status of remote populations, facilitate access to health information, and promote health education (Secinaro et al., 2021; Amjad et al., 2023; Santus et al., 2021).

The Economic-Industrial Health Complex (CEIS) also plays a crucial role in the development of the Global South. Through collaboration between companies, universities, research centers, and governments, CEIS can drive health technology research and development tailored to the region's specific needs, generate employment and income, and strengthen health infrastructure (Jakovljevic et al., 2021; Gadelha et al., 2015; Metten et al., 2015).

However, to fully harness the potential of technology, Global South nations must build their own digital and health sovereignty. Digital sovereignty ensures control over data and digital infrastructure, protecting the privacy and security of health information. Health sovereignty allows countries to make autonomous decisions about health policies, drug production, and technology access.

Building a healthier and more equitable future for the Global South requires a joint commitment between governments, institutions, and civil society. It is necessary to invest in robust and resilient health systems, promote health education, and ensure universal access to quality health services. Responsibly and ethically used technology can be a powerful tool to reduce health disparities and build a more just and healthy world for all.

Global South nations must take the lead in shaping their own health future, seizing technological opportunities and combating persistent disparities. Through South-South cooperation and the construction of digital and health sovereignty, it is possible to chart a path toward a healthier, more prosperous, and equitable future for all.

Artificial Intelligence in Healthcare as a Transformational Catalyst

AI emerges as a transformative catalyst in the global health landscape, opening up possibilities to overcome persistent challenges and build a healthier future for populations worldwide. For Global South countries, AI represents a unique opportunity to develop more efficient, equitable, and innovative health systems.

AI has immense potential to revolutionize disease diagnosis and treatment. Machine learning algorithms can analyze large volumes of medical data, such as images, health records, and lab tests, with greater accuracy and speed than human capability. This enables early disease detection, personalized treatments, and resource management optimization (Cordeiro et al., 2016, 2017; Noorbakhsh-Sabet et al., 2019; Paton & Kobayashi, 2019; Dias & Torkamani, 2019; Barbosa et al., 2021; Gomes et al., 2020, 2021, 2023, 2023b; de Oliveira et al., 2020; da Silva et al., 2021; Fonseca et al., 2022). A notable example is China's AI-powered system for lung cancer diagnosis, which surpasses the accuracy of human radiologists (Boeken et al., 2023).

AI is also paving the way for the accelerated development of new drugs and vaccines. By analyzing massive datasets of molecules and proteins, AI can identify new compounds with therapeutic potential and aid in creating personalized vaccines to combat specific diseases (Kannan et al., 2020). This capability can revolutionize medicine, especially in tackling neglected diseases and finding solutions for complex diseases. For instance, Vietnam is using AI to develop new vaccines against infectious diseases such as dengue and chikungunya (Yang et al., 2023).

AI can significantly expand access to quality healthcare, especially for populations in remote or resource-limited areas. AI-powered telemedicine allows patients to consult doctors remotely, receive diagnoses, and follow-up care, reducing the need for travel and facilitating access to specialized care. India, for example, is implementing AI-driven telemedicine programs to meet the health needs of rural communities (Tyagi & Sripriya, 2015; Haleem et al., 2021).

AI can be a powerful tool for strengthening health surveillance and combating pandemics. By analyzing data from various sources, such as social networks, medical records, and geolocation data, AI can identify disease spread patterns, predict outbreaks, and aid in decision-making to contain infectious diseases (Nain et al., 2021; Iqbal et al., 2021). During the Covid-19 pandemic, countries like Brazil and South Africa used AI to monitor the virus's spread and take disease control measures (Nain et al., 2021; Iqbal et al., 2021).

Despite AI's immense potential in healthcare, it is important to recognize the challenges and ethical considerations that accompany this technology. Ensuring the safety and reliability of AI systems, protecting health data privacy, and mitigating algorithmic bias are crucial aspects for the responsible implementation of AI in healthcare. Creating regulatory frameworks and promoting education in AI ethics are essential measures to ensure the responsible use of this technology in the healthcare sector.

AI stands as a transformational catalyst in healthcare, with the potential to revolutionize diagnosis, treatment, drug development, and access to quality healthcare. For Global South countries, AI represents a unique opportunity to develop more efficient, equitable, and innovative health systems, contributing to a healthier future for their populations. Responsible AI use, coupled with international cooperation and investment in research and development, will enable Global South countries to fully benefit from this technology's transformative potential, building a more prosperous and equitable future for all.

The Economic-Industrial Health Complex: A Pillar of Autonomous Development

In a global scenario marked by public health challenges, such as emerging pandemics, non-communicable chronic diseases, and persistent disparities in access to quality healthcare, the Economic-Industrial Health Complex (*Complexo Econômico-Industrial da Saúde* in Portuguese [CEIS]) emerges as a fundamental pillar for autonomous development and the achievement of health sovereignty for Global South countries (Pereira Filho et al., 2020; Miranda et al., 2019; Almeida et al., 2010). Inspired by thinkers like Carlos Gadelha, Sérgio Arouca, and Celso Furtado, we can understand CEIS as a strategic tool for building a healthier and more prosperous future for the populations of the Global South (Arouca, 2003; Gadelha & Costa, 2012; Gadelha et al., 2011, 2012, 2012b; Gadelha, 2006, 2022; Furtado, 2019).

CEIS is crucial for strengthening the health sovereignty of Global South countries, allowing them to make autonomous decisions about their health policies, drug production, and access to technologies. In the context of growing global health challenges, such as the Covid-19 pandemic, dependence on imports of medicines and medical supplies can pose a risk to national security and the ability to respond to health crises (Pereira Filho et al., 2020; Miranda et al., 2019; Almeida et al., 2010). Developing a robust CEIS enables Global South countries to reduce this external dependence, strengthen their national production capacity for health goods and services, and make strategic decisions about the health of their populations (Pereira Filho et al., 2020; Miranda et al., 2019; Almeida et al., 2010).

CEIS acts as a catalyst for technological innovation in health, driving the development of new products, services, and solutions to public health challenges (Pereira Filho et al., 2020; Miranda et al., 2019; Almeida et al., 2010). The integration of research, development, and national production of health goods and services allows Global South countries to adapt solutions to their specific needs, considering sociocultural, epidemiological, and resource characteristics (Pereira Filho et al., 2020; Miranda et al., 2019; Almeida et al., 2010). This adaptive capacity is crucial for developing innovative and effective solutions to major health challenges affecting Global South populations, such as neglected diseases, infectious diseases, and chronic diseases.

CEIS can be an engine of inclusive economic growth, generating employment, income, and opportunities for local development (Pereira Filho et al., 2020; Miranda et al., 2019; Almeida et al., 2010). National production of health goods and services boosts the pharmaceutical industry, medical equipment industry, and related sectors, creating new job opportunities and contributing to overall economic development (Pereira Filho et al., 2020; Miranda et al., 2019; Almeida et al., 2010). Additionally, CEIS can help reduce social inequalities by promoting access to quality healthcare for marginalized populations and reducing dependence on imported medicines and medical supplies (Pereira Filho et al., 2020; Miranda et al., 2019; Almeida et al., 2010).

However, developing CEIS in Global South countries faces challenges such as the need for investment in research and development, training qualified human resources, creating a regulatory environment conducive to innovation, and promoting public-private sector collaboration. Governments in Global South countries must implement public policies that encourage CEIS development, promote technology transfer, and build national capacities in science, technology, and innovation (Pereira Filho et al., 2020; Miranda et al., 2019; Almeida et al., 2010). Additionally, it is crucial to strengthen research institutions, universities, and national companies to contribute to developing a robust and sustainable CEIS.

China and Vietnam stand out as inspiring examples of CEIS development in the Global South (Kong et al., 2023; Pan et al., 2016; Sun et al., 2008; Chan & Daim, 2018; Chen et al., 2019; Sang & An, 2023; Sturgeon & Zylberberg, 2016; Thu et al., 2023; Kosowicz et al., 2023). China, for example, has invested heavily in pharmaceutical research and development, becoming a global hub for producing generic and innovative drugs (Kong et al., 2023; Pan et al., 2016; Sun et al., 2008; Chan & Daim, 2018). Vietnam, in turn, has excelled in developing telemedicine and AI technologies for health, expanding access to quality healthcare for its population, especially in rural and remote areas (Chen et al., 2019; Sang & An, 2023; Sturgeon & Zylberberg, 2016; Thu et al., 2023; Kosowicz et al., 2023).

Inspired by the examples of China and Vietnam, and guided by the reflections of thinkers like Gadelha, Arouca, and Furtado, Global South countries can build a future where health is not a privilege but a right that is accessible to all. South-

South collaboration, knowledge and experience sharing, and joint investment in research and development are crucial for Global South countries to achieve this goal (Pereira Filho et al., 2020; Miranda et al., 2019; Almeida et al., 2010).

The Economic-Industrial Health Complex, therefore, stands as a fundamental pillar for autonomous development and the achievement of health sovereignty for Global South countries (Pereira Filho et al., 2020; Miranda et al., 2019; Almeida et al., 2010). By strengthening health sovereignty, driving technological innovation, and generating inclusive economic growth, CEIS can contribute to building a healthier and more prosperous future for Global South populations (Pereira Filho et al., 2020; Miranda et al., 2019; Almeida et al., 2010). By developing a robust and sustainable CEIS, these countries can not only strengthen their health autonomy but also become global leaders in innovation and health solutions for the challenges of the 21st century.

Digital Colonialism and Imperialism

In an increasingly digitalized world, where information and communication have become essential pillars of social, economic, and political life, the issue of digital sovereignty emerges as crucial for the self-determination of peoples and the construction of more just and equitable societies. However, the rise of digital colonialism, a new form of exploitation and domination manifested through digital technologies, jeopardizes the digital sovereignty of Global South countries and deepens global inequalities.

Frantz Fanon's reflections on colonialism and imperialism, in works like *The Wretched of the Earth* and *Discourse on Colonialism*, offer crucial insights into understanding the nature and impacts of digital colonialism in the 21st century (Hilton, 2011; Fanon, 1980, 2015, 2022; Mbembe, 2021). Fanon, in analyzing the mechanisms of exploitation and domination used by colonial powers, reveals how control of information and culture becomes a fundamental instrument for maintaining power and subjugating colonized peoples.

In the context of digital colonialism, we observe unsettling similarities to the colonial practices described by Fanon (Adam, 2019; Schneider, 2022). The imposition of digital platforms and algorithms created by Global North companies, the massive extraction of personal data from Global South users, and mass surveillance are examples of how digital colonialism operates in the digital age. In this sense, imperial powers, through the control of digital technologies, seek to maintain their political, economic, and cultural dominance over other nations, perpetuating unequal power relations and deepening global inequalities.

Digital Colonialism Manifests in the Following Ways

- ***Imposition of digital platforms and algorithms***: Global North companies dominate the digital platforms and online services market, imposing their standards, values, and culture on Global South users. This limits the capacity of Global South countries to develop their own technological solutions and control access to information within their territories.

- ***Extraction of personal data***: Large technology companies from the Global North collect vast amounts of personal data from Global South users without proper consent and transparency about how this data is used. This practice raises concerns about data privacy and the use of this data for commercial or political purposes.
- ***Mass surveillance***: Governments and companies use mass surveillance technologies to monitor citizens' online activities, often without their knowledge or consent. This practice violates privacy and freedom of expression rights and can be used to suppress political and social dissent.

Main impacts of Digital Colonialism

- ***Loss of digital sovereignty***: Dependence on digital technologies created by Global North companies limits Global South countries' ability to control their data, protect their privacy, and develop their technological solutions. This makes them more vulnerable to exploitation and manipulation by foreign powers.
- ***Deepening global inequalities***: Digital colonialism contributes to deepening global inequalities by concentrating power and resources in the hands of a small number of companies and nations from the Global North. This limits the development opportunities for Global South countries and increases their technological dependence.
- ***Threat to democracy and human rights***: Digital colonialism can be used to suppress political and social dissent, manipulate elections, and violate human rights. This represents a serious threat to democracy and the rule of law in Global South countries.

To overcome the challenges of digital colonialism and build a more just and equitable digital future, it is crucial for Global South nations to:

- ***Strengthen their national states***: Building strong states with economic planning capacity and robust public policies is crucial to facing the challenges of digital colonialism. These states should be supported by a strengthened civil society capable of exercising social control and pushing for public policies that meet the population's needs.
- ***Invest in digital literacy***: Equipping the population with digital skills is essential to stimulate innovation, entrepreneurship, and critical use of information and communication technologies (ICTs). This includes developing formal and informal education programs that cover basic to advanced skills in areas like programming, data analysis, and digital security.
- ***Promote local digital culture***: Encouraging the production and dissemination of local digital content is crucial to strengthening cultural identity and providing alternatives to the dominant narratives from the Global North. This includes supporting local audio-visual production initiatives, developing video games

and educational games, and creating content-sharing platforms that allow the diffusion of local narratives and knowledge.
- ***Develop public policies for the digital sector***: Governments in the Global South must implement public policies that encourage the development of the local digital industry, social inclusion in ICT access, and data privacy protection. Examples of such public policies include investing in technology startups, creating subsidies for research and development of national technological solutions, promoting net neutrality, and implementing robust personal data protection laws.
- ***Strengthen global digital governance***: Global South countries should have an active voice in formulating rules and governance of the internet to ensure an inclusive and fair digital environment that respects national sovereignty. This involves active participation in international forums on digital governance, defending principles like multilateralism and multi-stakeholderism, and promoting South-South cooperation to build a more balanced global digital space.

Overcoming digital colonialism involves more than passive resistance to domination practices imposed by foreign powers. It requires building a digitally autonomous Global South capable of innovating and developing its technological solutions to meet specific needs. This requires investment in research and development, creating technological innovation hubs, encouraging digital entrepreneurship, and training highly qualified workers in science, technology, engineering, and mathematics (STEM).

Digital colonialism is a complex and multifaceted challenge that requires a strategic response from Global South countries. By recognizing the similarities with past colonial practices and drawing inspiration from the reflections of Frantz Fanon, we can build a more just and equitable digital future. The fight for digital sovereignty and self-determination in the digital world is crucial for building more prosperous, inclusive, and resilient societies in the Global South. Through strengthening states, investing in digital education and culture, implementing inclusive public policies, and actively participating in global digital governance, Global South countries can become protagonists in the new digital era.

Digital and Health Sovereignty

Digital sovereignty is crucial for the self-determination of peoples and the construction of more just and equitable societies in the digital age. It refers to the ability of a country or region to control its digital space, including infrastructure, data, platforms, and technologies operating within its territory. This means having the autonomy to:

- ***Define its own technology-related laws and policies***: The country must have the freedom to determine how digital technologies are used within its territory, including personal data protection, citizens' privacy, and online freedom of expression.

- ***Develop and implement its own technological solutions*:** Not depending on technologies created by other countries but having the ability to develop solutions that meet specific needs and are compatible with its values and culture.
- ***Protect its data and digital infrastructure*:** Ensuring the security and integrity of its data, preventing the misuse of information by third parties.

For Global South countries, digital sovereignty is even more important in a context marked by power asymmetries and increasing dependence on technologies created by Global North nations. Achieving digital sovereignty in this context is fundamental for:

- ***Promoting economic and social development*:** The ability to develop and implement home-grown technological solutions can drive economic growth, job creation, and innovation across various sectors.
- ***Reducing inequalities*:** Digital sovereignty can help reduce social and economic inequalities by ensuring that all citizens have equitable access to digital technologies and their benefits.
- ***Protecting culture and identity*:** Controlling their digital space allows Global South countries to protect their culture and identity, avoiding the imposition of external values and narratives.
- ***Strengthening democracy and human rights*:** Digital sovereignty is essential to guarantee freedom of expression, citizens' privacy, and social control over digital technologies.

Artificial Intelligence (AI) is one of the most disruptive technologies of the digital age, with the potential to transform various sectors, from healthcare to agriculture. However, the concentration of AI development and application in Global North countries poses a serious challenge to the digital sovereignty of the Global South.

To ensure that AI benefits their citizens, Global South nations need to become AI sovereign. This means:

- ***Investing in AI research and development*:** Allocating financial and human resources to develop AI solutions tailored to the specific needs and challenges of Global South countries.
- ***Creating their own AI platforms and algorithms*:** Developing AI platforms that respect the values and culture of Global South countries and are not controlled by foreign companies or governments.
- ***Training a skilled AI workforce*:** Investing in education and training professionals specialized in AI to ensure the capacity to develop and implement AI solutions in Global South countries.

Data has become a strategic resource in the digital age, with invaluable value for developing technologies like AI. However, the improper collection and use of personal data can pose a serious threat to citizens' privacy and the digital sovereignty of countries. To ensure digital sovereignty in this context, it is essential for Global South countries to:

- ***Implement robust personal data protection laws*:** Ensure that citizens' data is collected and used transparently, with explicit consent and in accordance with national laws.
- ***Develop infrastructure for data storage and processing*:** Create their own infrastructure to store and process citizens' data, avoiding reliance on foreign cloud storage services.
- ***Promote digital literacy*:** Educate citizens about the importance of personal data protection and how to use digital technologies safely and responsibly.

Unlike oil, data has unique characteristics that distinguish it from traditional commodities. Data is dynamic and renewable, with the potential to be constantly generated and improved. Additionally, data has contextual and strategic value that transcends immediate monetary value. Through data analysis and processing, valuable insights can be extracted for developing innovative solutions in various fields of knowledge.

The sale of raw data from the Global South to Global North nations, without proper protection and valorization, can lead to several negative consequences:

- ***Loss of control over strategic data*:** The Global South loses the ability to use its data to develop solutions that address its specific needs and challenges.
- ***Deepening technological dependence*:** Dependence on AI technologies and solutions developed by Global North countries increases the vulnerability of the Global South and limits its autonomy in the digital age.
- ***Brain drain*:** The lack of opportunities to develop AI solutions in the Global South can lead to the migration of talent to countries with better infrastructure and investment in research and development, widening the technological gap between regions.

Instead of focusing on the mere export of raw data, the Global South should seek alternatives that promote the development of a robust and sustainable data economy. This means:

- ***Investing in infrastructure for data storage and processing*:** Creating their own infrastructure to store and process citizens' data, avoiding reliance on foreign cloud storage services.
- ***Developing capabilities in data analysis and AI*:** Investing in the training of professionals specialized in data analysis and AI to ensure the capacity to develop and implement AI solutions in Global South countries.
- ***Creating regulatory frameworks for data protection*:** Ensuring transparency, citizen control over their data, and the ethical and responsible use of data.
- ***Promoting South-South cooperation*:** Encouraging collaboration between Global South countries to share knowledge, experiences, and resources in the pursuit of joint solutions for digital age challenges.

By transforming data into knowledge and innovation, the Global South can use data as a powerful instrument for sustainable economic and social development. Building a robust data economy, focusing on valorization and strategic use of data, can lead to a more autonomous, just, and prosperous digital future for its citizens.

Coupled with this, in a world increasingly challenged by health crises, health sovereignty emerges as a crucial concept for the self-determination of peoples and the construction of more just and equitable health systems. For the Global South, achieving public health autonomy is a fundamental challenge, especially given power asymmetries and dependence on foreign technologies and medicines.

In this context, digital sovereignty presents itself as a powerful platform to strengthen the health sovereignty of the Global South, driving the development of innovative and effective solutions for major public health challenges. The strategic application of AI in health can:

- ***Enhance epidemiological surveillance*:** Monitor and track disease spread in real-time, enabling faster and more effective responses to epidemics and pandemics.
- ***Analyze large-scale health data*:** Identify patterns and trends that can aid in disease prevention, developing more targeted public policies, and optimizing health system management.
- ***Develop more accurate diagnostic systems*:** Use AI algorithms to perform faster and more accurate diagnoses, reducing patient wait times and the risk of medical errors.
- ***Personalize disease treatment*:** Create individualized treatment plans for each patient, considering their genetic characteristics, medical history, and lifestyle, known as precision health.
- ***Develop new medicines and vaccines*:** Accelerate the research and development process of new medicines and vaccines, combating neglected diseases and responding to new public health threats.

For AI in health to reach its full potential and contribute to strengthening the health sovereignty of the Global South, it is essential to develop and strengthen the national economic-industrial health complex. This means:

- ***Investing in health technology research and development*:** Encouraging and supporting the research and development of national technologies in areas such as biotechnology, pharmaceuticals, medical devices, and health software.
- ***Creating and strengthening public health enterprises*:** Establishing robust and efficient public enterprises to produce medicines, vaccines, medical devices, and other essential health products and services.
- ***Promoting the national IT and AI industry*:** Fostering the development of the national IT and AI industry, focusing on creating technological solutions for the health sector.

- ***Encouraging public-private partnerships***: Establishing strategic partnerships between the public and private sectors for the joint development of health technologies and solutions, leveraging the capabilities and resources of both sectors.

In the current configuration, AI and national sovereignty are in a symbiotic relationship: developing the national health industrial complex, driven by AI, is crucial for achieving health sovereignty while enabling the self-determination of Global South countries with digital sovereignty. Through the national production of medicines, vaccines, medical devices, and health software, countries can reduce their dependence on technological imports while strengthening their ability to respond to health crises.

AI, therefore, can be a powerful instrument to optimize the production and distribution of health products and services, improve the quality of healthcare, and drive innovation in the sector. By investing in AI research and development in healthcare, Global South countries can develop innovative and personalized solutions to their specific needs, contributing to building more just and equitable health systems.

Pursuing health sovereignty through AI and developing the national health industrial complex is fundamental for the Global South to achieve sustainable development and ensure the well-being of its population. Through autonomy in healthcare, countries can strengthen their capacity to handle health crises, reduce health inequalities, and build a healthier and more prosperous future for all citizens.

Investing in research, development, and innovation in healthcare, together with promoting the national industry and building strategic partnerships, is crucial for the Global South to take control of its destiny in healthcare and build a more autonomous, resilient, and just future for its citizens.

Challenges and Opportunities

AI presents itself as a powerful tool to transform the health sector, driving innovation, optimizing processes, and improving the quality of healthcare. However, the application of AI in health also presents challenges and opportunities that need careful consideration, especially in the context of digital imperialism and the quest for health sovereignty in the Global South.

Challenges of AI in Health

1. **Unequal Access to Technology:** The concentration of AI development and application in Global North countries can deepen health inequalities, limiting access for Global South countries to the benefits of the technology.
2. **Ethical and Regulatory Issues:** The application of AI in health raises complex ethical issues, such as the use of biased algorithms, the protection of

personal data, and the role of human intelligence in medical decision-making. Robust regulatory frameworks adapted to the reality of the Global South are crucial to ensuring the ethical and responsible use of AI in health.

Opportunities of AI in Health for the Global South

1. **Strengthening Health Sovereignty:** AI can be a powerful tool for strengthening the health sovereignty of the Global South, enabling the development of national solutions for public health challenges, reducing dependence on foreign technologies, and promoting self-determination in health.
2. **Improving the Quality of Healthcare:** AI can contribute to improving the quality of healthcare through more accurate and early diagnosis of diseases, personalized treatment, and optimized management of health systems.
3. **Developing Innovative Solutions:** The Global South has the opportunity to use AI to develop innovative and personalized solutions for its specific health challenges, taking into account the characteristics and needs of its populations.
4. **Promoting Health Equity:** AI can be used to promote health equity by reducing disparities in access to quality healthcare and ensuring that all citizens benefit from the technology.

Through a joint and coordinated effort, the Global South can overcome the challenges of digital imperialism and build a more just, equitable, and resilient health future where AI is used as a tool for the well-being of the population and the strengthening of health sovereignty. Economic blocs like BRICS+ can contribute to the mutual development of Global South nations through fair agreements that help overcome the dependency logic built by yesterday's and today's colonial powers.

Conclusion

The pursuit of digital and health sovereignty is a strategic imperative for the Global South in a world increasingly interconnected and challenged by health crises, power asymmetries, and inequalities. Artificial Intelligence (AI) emerges as a powerful tool to drive sustainable development, strengthen the autonomy of countries, and build a more prosperous and just future for their citizens.

By investing in AI research, development, and innovation, along with promoting the national technology industry and building strategic partnerships, the Global South can take control of its destiny in the digital era and build a more autonomous, resilient, and just future. However, to fully benefit from AI, the Global South must overcome the challenges of digital imperialism and pave the way for digital and health sovereignty.

Achieving digital and health sovereignty is a complex and challenging historical process that requires firm political commitment from responsible historical blocs, leading strong national states with broad participation from organized civil society, strategic investments, and South-South cooperation. By developing their technological capabilities, becoming sovereign in AI, and gaining control over their data, the Global South can transform the challenges of digital colonialism into opportunities for sustainable development, social inclusion, and the construction of more just and equitable societies. The struggle for digital and health sovereignty is not only a technical issue but also a political and social one. It is the fight for self-determination in the digital age, the ability to define their future, and to build societies based on values such as democracy, justice, and solidarity.

The Global South has the opportunity to become a protagonist in the AI era, using this powerful technology to build a more prosperous and just future for its citizens. Through collaboration, innovation, and determination, we can build a world where AI is used for the well-being of humanity and the sustainable development of the planet.

References

Adam, T. (2019). Digital neocolonialism and massive open online courses (MOOCs): Colonial pasts and neoliberal futures. *Learning, Media, and Technology*, 44(3), 365–380.

Almeida, C., Campos, R.P.D., Buss, P., Ferreira, J.R., & Fonseca, L.E. (2010, March). A concepção brasileira de cooperação Sul-Sul estruturante em saúde. *R. Eletr. de Com. Inf. Inov. Saúde. Rio de Janeiro*, 4(1), 25–35.

Amjad, A., Kordel, P., & Fernandes, G. (2023). A review on innovation in healthcare sector (telehealth) through artificial intelligence. *Sustainability*, 15(8), 6655.

Arouca, S. (2003). *O dilema preventivista*. São Paulo: UNESP, 1975.

Barbosa, V.A.F., Gomes, J.C., de Santana, M.A., Albuquerque, J.E.D.A., de Souza, R.G., de Souza, R.E., & dos Santos, W.P. (2021). Heg. IA: An intelligent system to support diagnosis of Covid-19 based on blood tests. *Research on Biomedical Engineering*, 1–18.

Boeken, T., Feydy, J., Lecler, A., Soyer, P., Feydy, A., Barat, M., & Duron, L. (2023). Artificial intelligence in diagnostic and interventional radiology: Where are we now? *Diagnostic and Interventional Imaging*, 104(1), 1–5.

Carter, J., Friedland, J.S., Kirwan, D.E., & Nathavitharana, R.R. (2020). Translating scientific discoveries during pandemics: Ensuring equity for people affected by Covid-19 and tuberculosis. *ERJ Open Research*, 6(4).

Chan, L., & Daim, T. (2018). A research and development decision model for pharmaceutical industry: Case of China. *R&D Management*, 48(2), 223–242.

Chen, X., Xue, S., Lv, M., & Wang, R. (2019). Pharmaceutical industry in China: Policy, market and IP. *In*: In: K.C. Liu, U. Racherla (Eds.), *Innovation, Economic Development, and Intellectual Property in India and China* (pp. 215-250). ARCIALA Series on Intellectual Assets and Law in Asia. Springer, Singapore.

Cordeiro, F.R., Santos, W.P., & Silva-Filho, A.G. (2016). An adaptive semi-supervised Fuzzy GrowCut algorithm to segment masses of regions of interest of mammographic images. *Applied Soft Computing*, 46, 613–628.
Cordeiro, F.R., Santos, W.P.D., & Silva-Filho, A.G. (2017). Analysis of supervised and semi-supervised GrowCut applied to segmentation of masses in mammography images. *Computer Methods in Biomechanics and Biomedical Engineering: Imaging & Visualization*, 5(4), 297–315.
da Silva, C.C., de Lima, C.L., da Silva, A.C.G., Silva, E.L., Marques, G.S., de Araújo, L.J.B., ... & da Silva Filho, A.G. (2021). Covid-19 dynamic monitoring and real-time spatio-temporal forecasting. *Frontiers in Public Health*, 9, 641253.
de Oliveira, A.P.S., de Santana, M.A., Andrade, M.K.S., Gomes, J.C., Rodrigues, M.C., & dos Santos, W.P. (2020). Early diagnosis of Parkinson's disease using EEG, machine learning, and partial directed coherence. *Research on Biomedical Engineering*, 36, 311–331.
Dias, R., & Torkamani, A. (2019). Artificial intelligence in clinical and genomic diagnostics. *Genome Medicine*, 11(1), 70.
Fanon, F. (1980). *Em Defesa da Revolução Africana* (Pascoal, I., trad.). Lisboa: Sá da Costa. (Original publicado em 1969).
Fanon, F. (2015). *Peau noire, masques blancs*. Média Diffusion.
Fanon, F. (2022). *Os condenados da terra*. Editora Schwarcz-Companhia das Letras.
Fonseca, F.S., Torcate, A.S., Da Silva, A.C.G., Freire, V.H.W., De Farias, G.P.D.M., De Oliveira, J.F.L., ... & Dos Santos, W.P. (2022, November). Early prediction of generalized infection in intensive care units from clinical data: A committee-based machine learning approach. *In*: *2022 IEEE Latin American Conference on Computational Intelligence* (*LA-CCI*) (pp. 1– 6). IEEE.
Furtado, C. (2019). Os desafios da nova geração. *Brazilian Journal of Political Economy*, 24, 483–486.
Gadelha, C.A.G. (2006). Desenvolvimento, complexo industrial da saúde e política industrial. *Revista de Saúde* Pública, 40, 11–23.
Gadelha, C.A.G. (2022). Complexo Econômico-Industrial da Saúde: A base econômica e material do Sistema Único de Saúde. *Cadernos de Saúde Pública*, 38, e00263321.
Gadelha, C.A.G., Machado, C.V., Lima, L.D.D., & Baptista, T.W.D.F. (2011). Saúde e territorialização na perspectiva do desenvolvimento. *Ciência & Saúde Coletiva*, 16, 3003–3016.
Gadelha, C.A.G., & Costa, L.S. (2012). Saúde e desenvolvimento no Brasil: avanços e desafios. *Revista de Saúde Pública*, 46, 13–20.
Gadelha, C.A.G., Costa, L.S., Borges, T.R., & Maldonado, J.M.S.D.V. (2012). O complexo econômico-industrial da saúde: Elementos para uma articulação virtuosa entre saúde e desenvolvimento. *Saúde em Debate*, 36(92), 21–30.
Gadelha, C.A.G., Costa, L.S., & Maldonado, J. (2012b). O complexo econômico-industrial da saúde e a dimensão social e econômica do desenvolvimento. *Revista de Saúde Pública*, 46, 21–28.
Gadelha, C.A., Costa, L.S., & Maldonado, J. (2015). The economic-industrial health care complex and the social and economic dimension of development. *Health Innovation Systems, Equity and Development*, 20.
Gomes, J.C., Barbosa, V.A.D.F., Santana, M.A., Bandeira, J., Valença, M.J.S., de Souza, R.E., ... & dos Santos, W.P. (2020). IKONOS: An intelligent tool to support diagnosis of Covid-19 by texture analysis of X-ray images. *Research on Biomedical Engineering*, 2022, 38(1), 1–14.

Gomes, J.C., de Freitas Barbosa, V.A., de Santana, M.A., de Lima, C.L., Calado, R.B., Júnior, C.R.B., ... & dos Santos, W.P. (2023). Rapid protocols to support Covid-19 clinical diagnosis based on hematological parameters. *Research on Biomedical Engineering*, 39(3), 509–539.

Gomes, J.C., de Santana, M.A., Masood, A.I., de Lima, C.L., & dos Santos, W.P. (2023b). Covid-19's influence on cardiac function: A machine learning perspective on ECG analysis. *Medical & Biological Engineering & Computing*, 61(5), 1057–1081.

Guzmán, C.A.F. (2022). Complexity in global health: Bridging theory and practice. *Annals of Global Health*, 88(1).

Haleem, A., Javaid, M., Singh, R.P., & Suman, R. (2021). Telemedicine for healthcare: Capabilities, features, barriers, and applications. *Sensors International*, 2, 100117.

Hilton, B.T. (2011). Frantz Fanon and colonialism: A psychology of oppression. *Journal of Scientific Psychology*, 12(1), 45–59.

Ho, A., & Dascalu, I. (2020). Global disparity and solidarity in a pandemic. *Hastings Center Report*, 50(3), 65–67.

Iqbal, S., Ahmad, S., Bano, B., Akkour, K., Alghamdi, M.A.A., & Alothri, A.M. (2021). A systematic review: Role of artificial intelligence during the Covid-19 pandemic in the healthcare system. *International Journal of Intelligent Information Technologies (IJIIT)*, 17(1), 1–18.

Islam, M.R. (2023). Navigating the health and wellbeing landscape in South Asia: Assessing progress, addressing challenges, and charting a path forward: Navigating the health and wellbeing landscape in South Asia. *Eastern Journal of Healthcare*, 3(1), 8–17.

Jain, P. (2021). Impact of COVID-19 pandemic on global healthcare systems and the role of a new era of global collaborations. *Sushruta Journal of Health Policy & Opinion*, 14(3), 1–5.

Jakovljevic, M., Liu, Y., Cerda, A., Simonyan, M., Correia, T., Mariita, R.M., ... & Varjacic, M. (2021). The Global South political economy of health financing and spending landscape: History and presence. *Journal of Medical Economics*, 24(sup1), 25–33.

Kannan, S., Subbaram, K., Ali, S., & Kannan, H. (2020). The role of artificial intelligence and machine learning techniques: Race for Covid-19 vaccine. *Archives of Clinical Infectious Diseases*, 15(2).

Kasai, T. (2019). For the future: Delivering better health in the Western Pacific region. *Journal of Global Health Science*, 1(1).

Kong, L., Li, Q., Kaitin, K.I., & Shao, L. (2023). Innovation in the Chinese pharmaceutical industry. *Nature Reviews Drug Discovery*, 22(1), 12–13.

Kosowicz, L., Tran, K., Khanh, T.T., Dang, T.H., Pham, V.A., Ta Thi Kim, H., ... & Nguyen, T.A. (2023). Lessons for Vietnam on the use of digital technologies to support patient-centered care in low- and middle-income countries in the Asia-Pacific region: Scoping review. *Journal of Medical Internet Research*, 25, e43224.

Loewenson, R., Villar, E., Baru, R., & Marten, R. (2021). Engaging globally with how to achieve healthy societies: Insights from India, Latin America and East and Southern Africa. *BMJ Global Health*, 6(4), e005257.

Mabhala, M.A., Yohannes, A., Massey, A., & Reid, J.A. (2020). Mind your language: Discursive practices produce unequal power and control over infectious disease: A critical discourse analysis. *International Journal of Preventive Medicine*, 11.

Mbembe, A. (2012). Metamorphic thought: The works of Frantz Fanon. *African Studies*, 71(1), 19–28.

Metten, A., Costa, L.S., Gadelha, C.A.G., & Maldonado, J. (2015). The introduction of the health economic industrial complex in the development agenda: An analysis from Kingdon's multiple streams model. *Revista de Administração Pública*, 49(4), 915.

Miranda, J., Neto, J., Magalhães, J., & Pessoa, V. (2019). Desafios da saúde pública no século XXI. *Anais do Instituto de Higiene e Medicina Tropical*, 18, 134–137.

Nain, M., Sharma, S., & Chaurasia, S. (2021). Pandemic management using artificial intelligence-based safety measures: Prediction and prevention. *In*: *Machine Learning and Data Analytics for Predicting, Managing, and Monitoring Disease* (pp. 86–110). IGI Global.

Nguyen, T.L.H. (2022). On Improving Healthcare with a world perspective: Evidence for Global Health Programs. *International Healthcare Review* (online), 1(1).

Noorbakhsh-Sabet, N., Zand, R., Zhang, Y., & Abedi, V. (2019). Artificial intelligence transforms the future of health care. *The American Journal of Medicine*, 132(7), 795–801.

Pan, Y., Huang, Y., & Qi, E. (2016). Chinese manufacturing industry development strategy from the financial perspective: Grey modeling and analysis of data from the listing corporation China pharmaceutical manufacturing enterprises. *Grey Systems: Theory and Application*, 6(2), 203–215.

Paton, C., & Kobayashi, S. (2019). An open science approach to artificial intelligence in healthcare. *Yearbook of Medical Informatics*, 28(01), 047–051.

Pereira Filho, J.E., dos Santos, R.R., de Bragança, J.C.O., & Rothstein, J.M.E.M. (2020). O complexo econômico industrial da saúde e os desafios estruturais em época de pandemia. *Confluências| Revista Interdisciplinar de Sociologia e Direito*, 22(2), 51–78.

Sang, H.T., & An, V.T.T. (2023). Making middle-power alignment work: Reinforcing Taiwan-Vietnam collaboration in the semiconductor industry. *International Journal of China Studies*, 14(2), 125–152.

Santus, E., Marino, N., Cirillo, D., Chersoni, E., Montagud, A., Santuccione Chadha, A., ... & Lindvall, C. (2021). Artificial intelligence-aided precision medicine for Covid-19: Strategic areas of research and development. *Journal of Medical Internet Research*, 23(3), e22453.

Scheerens, C., Ruyssen, I., Ray, S., De Sutter, A., Vanhove, W., Bekaert, E., ... & De Maeseneer, J. (2020). Tackling adverse health effects of climate change and migration through intersectoral capacity building in Sub-Saharan Africa. *BJGP Open*, 4(2).

Schneider, N. (2022). Governable stacks against digital colonialism. Triple C: Communication, capitalism & critique. *Open Access Journal for a Global Sustainable Information Society*, 20(1), 19–36.

Secinaro, S., Calandra, D., Secinaro, A., Muthurangu, V., & Biancone, P. (2021). The role of artificial intelligence in healthcare: A structured literature review. *BMC Medical Informatics and Decision Making*, 21, 1–23.

Sturgeon, T., & Zylberberg, E. (2016). The global information and communications technology industry: Where Vietnam fits in global value chains. *World Bank Policy Research Working Paper* (7916).

Sun, Q., Santoro, M.A., Meng, Q., Liu, C., & Eggleston, K. (2008). Pharmaceutical policy in China. *Health Affairs*, 27(4), 1042–1050.

Thu, T.N.D., Nguyen, Q.K., Taylor-Robinson, A.W., & Tu, D.T.N. (2023). Healthcare in Vietnam: Harnessing artificial intelligence and robotics to improve patient care outcomes. *Cureus*, 15(9).

Tyagi, R., & Sripriya, S. (2015). 3-level integrated telemedicine network for improving healthcare access to rural India. *GSTF Journal of Nursing and Health Care (JNHC)*, 2(1).

Vishwanathan, A. (2022). Covid-19 pandemic: A multilayered challenge to South Asia. *Towards Excellence*, 14(4).

Yang, H., Nguyen, T.N., & Chuang, T.W. (2023). An integrative explainable artificial intelligence approach to analyze fine-scale land-cover and land-use factors associated with spatial distributions of place of residence of reported dengue cases. *Tropical Medicine and Infectious Disease*, 8(4), 238.

4 | Artificial Intelligence and Human Rights Now

Luiz Vianna Sobrinho

We accept the challenge to propose a bioethical approach to the issues raised by the advent of information technologies in the practical life world from a Human Rights perspective. With this horizon, the endless characteristic of the continuous struggle for the defense and demarcation of the human becomes even more necessary now, given the characteristics of a set of technological objects that blur our ontological boundaries more than ever. In the long relationship of our development in symbiosis with our technological creations, we have never been so close to confusing ourselves or comparing ourselves with these objects as we are now. At this moment, what most defines and distinguishes us in the realm of living beings is continuously presented to the common human as the main characteristic of this technological object, confronting us with a shared existence: intelligence. We propose an expanded redefinition of Human Rights, aiming at the ontological guarantee of our dignity and freedom.

You are not 'machine-learn'. Men is what you are.
— Charles Chaplin[1]

*Email: praxiscomplexidade@gmail.com

[1] Author's note: The original quote by Charles Chaplin, in his "Final Speech" from the film *The Great Dictator*, says, "You are not machines! You are not cattle! You are men!" It is generally cited and translated by omitting the middle sentence, thus avoiding the focus on labor and accentuating the man-machine dualism. See [The Final Speech from *The Great Dictator*] https://www.charliechaplin.com/en/articles/29-the-final-speech-from-the-great-dictator.

For a Better Perception of the Zeitgeist[2]

More than half a century ago, Hannah Arendt, in one of her major works, provides a unique perspective on the dangers surrounding the human condition (Arendt, 2004). Her central thesis posits that thinking about this condition at this point in history was due to the possibility that it was under threat. In the preface, Arendt highlights the scientific efforts of her time to "make life itself artificial", pointing to three historical signs that led her to this hypothesis. The first was the desire to leave Earth, evident in the celebrations and media commentary following the launch and orbit of the first artificial satellite. The second was the aspiration to interfere with birth, through attempts to create life in a test tube, with the potential to "produce superior human beings"; and finally, the hope of extending life beyond one hundred years. These aspects of our life—birth, death, and our world—are the secular inheritances we have received, deeply marking what constitutes our condition, which we wish to replace with something "produced by ourselves" (Arendt, 2004, p. 10).

Today, the fundamental issue of birth, as initially addressed by the philosopher, seems trivial. Artificial insemination has become commonplace, and terms like "*in vitro* fertilization" or the euphemism "assisted reproduction" have made the method developed by science more widely accepted in our society. Advances in public health and the discovery of new pharmacological and therapeutic technologies in the latter half of the 20th century have catapulted life expectancy near the centenary mark in developed countries.

However, the quest for immortality or life beyond 100 years has led to questionable end-of-life medical practices (Camargos, 2017; Sales Júnior, 2017) and the application of dubious scientific methods regarding our death. Discussions are already moving beyond these issues, exploring genetic manipulation techniques that exceed safety limits (Cyranoski & Reardon, 2015) and intentionally revisiting the age-old topic of eugenics (Savulescu, 2001; Mukherjee, 2018). On a theoretical-philosophical level, as evidenced by its very term, 'transhumanism' already addresses another species, under different conditions (Bostrom, 2004; Vilaça & Palma, 2012).

Nonetheless, if we revisit Arendt's central argument at the time of writing, we will notice that it begins with her astonishment at the reaction and comments regarding the launch of the first artificial satellite placed in orbit by man. Leaving Earth became one of the main goals that science pointed to humanity, which, for Arendt, would shake one of the foundations of our human condition. The statement that alarmed her showed relief at the "first step to free man from his imprisonment on Earth". She noted that although the phrase was banal, it was

[2] Author's note: This text was partially published originally in 2019 in a collection of bioethics articles (Vianna Sobrinho, 2019, pp. 135–139). Interestingly, a similar allusion to Hannah Arendt's thoughts, linking them to Douglas Rushkoff's comments, appeared shortly thereafter with the same connotation in an important publication on ethics in artificial intelligence (Coeckelbergh, 2020).

extraordinary because "no one in the history of humanity had ever conceived of the Earth as a prison for the human body" (Arendt, 2004, p. 10).

Where did this desire of men, of the technoscience of the 20th century, to leave Earth go? After the initial decades filling our orbit with all kinds of satellites, manned missions to the moon, and robotic explorations to others, humanity has reached the limits of the solar system, mapping its anatomy. This achievement is often compared to the voyages of discovery; for some, it is a feat of such complexity that the layperson cannot grasp its magnitude (Sagan, 1996). But can we still say that humanity at large continues to aspire to leave Earth? Just as other scientific advances regarding birth and death, like *in vitro* fertilization and genetic manipulation or the use of oxidative molecules for a longer life, have come to fruition, can we still say that humans want to leave Earth? Have we given up on this desire? Or are we already leaving this world?

Sixty years after this symptom was announced by Arendt, media specialist and theorist Douglas Rushkoff tells us a story that would be terrifying for the philosopher, even if only for its possibility in the fictional imagination (Rushkoff, 2018). Invited to give a talk on the future of technology at a super-luxury resort, he was surprised to find himself speaking in a private room to only six multimillionaires from the financial sector. After introductory questions about available technologies, Rushkoff realized what this small audience wanted to know when one of them asked how they should prepare for "the event".

The "event", Rushkoff explains, was the euphemism used by his cynical interlocutors to describe the moment when, in this unsolvable world, "environmental disaster, social unrest, nuclear explosion, uncontrolled virus, or robot hackers will destroy everything". Besides practical operational concerns—such as the question of valueless digital money to pay guards or robots to protect their refuges—the central issue of his interview revolved around post-human utopia. There was no mention of the possibilities technology could offer in solving humanity's central issues; nothing about climate change, rising sea levels, mass migrations, global pandemics, or resource depletion. Interested in topics like the feasibility of colonizing Mars, halting and reversing aging processes, and the real possibility of transferring their minds to supercomputers, they wanted to know how to "entirely transcend the human condition". For this, the entire digital revolution and "the future of technology is about one thing: escape" (Rushkoff, 2018).

We then realize that more than controlling birth with artificial techniques or mastering aging processes to delay death at any cost, humanity has earlier fulfilled the promise of leaving Earth. Not as expected, in special suits, with powerful rockets, with immense energy costs to overcome a powerful attraction trying to bring us back, possible only for a few privileged ones; but in a new guise of reality, continuously and ubiquitously, using tiny devices that keep us connected and expanded to the sensation of universality, so that everyone remains in this space. This is how we left Earth; but, as Rushkoff warns, in total consonance with Hannah Arendt's text:

> It is less a vision of humanity's migration to a new state of being than a quest to transcend all that is human: body, interdependence, compassion, vulnerability, complexity. As technology philosophers have been pointing out for years, the transhumanist vision very easily reduces all reality to data, concluding that humans are merely information-processing objects (Rushkoff, 2018).

By telling this story, we intend to demonstrate the period and conditions under which the sociocultural transformations of the technological transition come to us at this turn of the century. It is important to realize that some form of choice of a new and, perhaps, expanded reality is at the root of the formation of a new *ethos*.

The philosopher Luciano Floridi has been consolidating his thoughts on this moment we live in for decades, aiming to define a new era to adjust to what everyone calls the "fourth industrial revolution". After a distant pre-history, our generation would be the one living the transition from History to 'Hyper-history', where Information and Communication Technologies (ICT) make "all the difference between who we were, who we are, and who we can be and become" (Floridi, 2014, p. 1). Characteristically, this new reality where we will live, named 'infosphere' by this author, will require new reflections and conceptualizations both in the philosophy of nature and in a new philosophical anthropology. Humanity, he believes, is heading towards a definitive entry into the digital age, under a prophecy that marks this era: "Only a society that lives hyper-historically can be threatened informationally, by a cyberattack. Only those who live by the digit may die by the digit" (Floridi, 2014, p. 4).

This book, however, is aimed at the general public. It condenses Floridi's work into more straightforward language than typically used in academia. To better reflect his argument, it is necessary to address his robust proposal for a new philosophy of information. In this philosophy, data will assume the role of arché of reality (Floridi, 2011). His ontology will structure an expanded ethics for the digital age, with clear Spinozist inspiration, featuring a bold revision of the status of moral agents and patients (Floridi, 2013). He treats all informational organisms as 'inforgs'. In his preface, he declares his intention to avoid discussing political issues. He then develops the grand structure of his model, heavily relying on logic.

We observe Floridi's careful justification throughout each chapter and argument. He logically opposes other ethical views, especially the ethics of virtues. We believe that perceiving the ethics of virtues as "intrinsically anthropocentric and individualistic" (Idem, p. 61) is a mistake from a meta-ethical standpoint. This perception confuses his data-anchored interpretation of reality with the moral choices humans make, which are not tied to the logical structure of reality. Human moral development expanded the limits of protection and care for the moral patient, characteristic of bioethics and environmental ethics. This expansion is not contained within the structure of natural world reality. Our moral concepts are embodied in social life forms, where we choose our moral bonds and, consequently, the possible forms of social practice (MacIntyre, 1976). This is the extremely political consequence of our ethical choices.

By explicitly distancing himself from this, privileging the logic of data as a mathesis to structure his ethics, it becomes possible to disperse moral responsibility across a hybrid system, distributed among multi-agents (Floridi, 2013)—as if governments, non-governmental organizations, political parties, economic groups, and companies were not the social presence of human actions with individual responsibility.

It is important to delineate this because political choices will underpin our future concerns in the social settlement of this powerful technology. Floridi presents a strong ethical proposition for ICTs, assuming from the preface of his volume that a certain "constructive liberalism" is at the core of his thinking (Idem, p. xvi).

In the Beginning was the '*Logos*'[3]

The proclaimed revolution of the information age firmly establishes itself over the past decade with the development of a set of technologies with significant societal impact, collectively termed Artificial Intelligence (AI). Information technologies, in the everyday life of citizens, have mostly been support mechanisms for managing electronic technologies. However, at the turn of the last decade, with the ubiquitous employment of AI, the total datafication of social grammar, and platformization as the organization of social relations, it has become clear that power and control over data are central issues. Alongside this, the exponential speed of progress in the computational field has not allowed enough time for legal norms to ensure adequate individual protection or regulate the political impact of the emerging inequities.

The "siren song" of technological progress always promises humanity benefits by solving its worldly problems. However, history, especially in modern times, has shown us that the possession and design of technological development are tied to control and the maintenance of power relations. Despite the tremendous technological advancements of the last century, we have tragically exacerbated the unequal distribution of socially produced wealth and left a significant portion of the global population without access to these essential resources, without even the minimum conditions for human dignity. In this context, AI is presented as the technology that will help us solve all global crises. Recognizing that this has always been the promise of other technological waves, we await something that will redeem us from ourselves.

[3] Author's note: The Christian apostle John opens his book, which forms part of the *New Testament of the Bible*, with a phrase of significant and unavoidable meanings: "In the beginning was the *Logos* (λόγος), and the Logos was with God, and the Logos was God." Most Bibles in the Portuguese language translate '*logos*' as '*verbo*' (verb), while the English translation uses the term 'word'." Logos is a word to which we attribute many meanings in philosophy, psychology, and theology.

From the outset, we must identify what belongs to the realm of discourse and the various overlapping interpretations of the technological object. The frequent and erroneous debate points to the polarities of a technophobic or technophilic stance related to AI. Attempting a neutral position focused solely on the purely objective characteristics of the technical object's mechanism is misleading. Viewing a knife, a sewing machine, or a therapeutic decision algorithm as isolated instruments, without considering their reproduction alongside humans and their social relations, is a significant and naive mistake.

Many assert, in a fallacy that has become commonplace, that technological development has always produced waves of conservative rejection throughout history, exaggerating its risks to humanity's future, and that these fears never materialized, allowing technoscientific development to follow its course of continuous progress. However, we argue that there are at least two strong objections to this argument.

The first point is that history exhausts itself in the present we inhabit and does not follow this trajectory as a natural law. This interpretation is possible only because of the evident bias of the "winner's narrative" that this argument carries. It does not account for the extinction of peoples, cultures, and histories. A clear example is the primarily technological confrontation between Europeans and Amerindians in recent centuries—a massacre among many others.

The second point is that if we do not acknowledge the extinction risk from our technologies, we must ignore our fear of self-destruction by nuclear arsenal, which still haunts global geopolitics. We must also overlook the risks of creating uncontrollable chimeras through genetic engineering, which motivated the Asilomar Conference (Mukherjee, 2018). More evidently, we must forget the entire current debate on the Anthropocene and climate disaster, a direct consequence of industrial technological escalation (Gabriel, 2018).

Therefore, understanding the human relationship with the technical object within the evolutionary conception of epiphylogenetics (Stiegler, 2014) does not support the notion of continuity that supposedly ensures the security of a chain of new discoveries. No perspective on the history of technique or technology obliges us to accept this deterministic view of inevitable progress that invalidates the fear of civilizational failure. On the contrary, some authors argue that the accelerated exploitation by mass industrialization of mnemotechnics, reproduced for capitalist investment, may be reaching a point of no return. This exploitation could lead to psychic technification, resulting in a crisis of individuation and temporality, pushing us towards a world without a future (Stiegler, 2019).

At this moment, we are no longer technologically transforming a force of nature, such as hydraulics, or developing mechanical arms that function similarly in various animals. Nor are we merely learning to harness other physical properties like electricity or nuclear energy, which we have come to control with desirable safety. We are not even dealing with biological matter, where we delve into the basic data of biochemistry that encodes the forms of living beings. Instead, we are now dealing with something that characterizes us as humans, placing our ontological status in any conception of the universe we may consider from the socio-historical moment we live in. We are projecting onto our technological creation the emulation of our cognitive abilities, our forms of reason, and even affection. Since the beginning of Western thought, starting with the ancient Greeks, what we have today still revolves around the understanding and conquest of the *logos*.

In understanding the forms of our reason, it is important to highlight the firm and secure defence of values tied to human characteristics in a recent document by the highest pontiff of the Catholic Church. Pope Francis, in celebration of the 57th World Day of Peace, published a letter titled "Artificial Intelligence and Peace" (Francis, 2024), where he enumerates the main risks of information technologies that threaten humanity as a whole, understood in its intrinsic dignity as a single family. Throughout his defence of human uniqueness, he highlights the core of our ontology, in the "inaccessible space that escapes any attempt at quantification" (Idem). Clearly a text that, openly inspired by the Word, descends to the reality of the incarnate world and warns of the same secular fears as Floridi's prophecy: that our end may also come through the *logos*.

A New Impetus for Bioethics

This moment expands into a worldview, with the consequent adoption of new epistemological approaches, as seen in the field of health practices. What we now call "data medicine" (Vianna Sobrinho, 2021) could have been named after an important statement by physician and scholar Eric Topol in his book *The Creative Destruction of Medicine* (Topol, 2013). A master and mentor of generations of cardiologists at the end of the last century, Topol proposes a total transformation of medical techniques—confessedly inspired by Schumpeter—with the following prophecy: "Medicine is about to go through its greatest shakeup in history [...] for the first time, we can digitize humans" (Idem, p. 13). This sentence clearly marks the datafication of medicine and heralds it as a leader and pioneer of precision medicine, maintaining a prolific production of both academic work and guidelines for this new era of medical practice with the advent of AI technologies (Topol, 2019).

The "digitalization of humans" utilizes the reading of objective functional data from remote biosensors, the genetic sequencing of each individual, the storage of diagnostic images, and the continuous exchange of information forming the health ecosystem. The advancement of AI technologies has accelerated so rapidly, particularly after the development of machine learning models, that it has resulted in a new direction for expectations. We end the same decade with the theoretical conclusion that, soon, we may no longer need doctors (Vianna Sobrinho, 2021). This transformation gradually undermines the autonomy of medical decision-making, reducing the role of subjective opinions and intuitions, and steering towards the fully objective quantification of the health/disease process.

As we argue, for this conception, the epistemological basis of medicine has practically moved towards an assumedly more naturalistic view, previously advocated by the biostatistical model of philosopher Christopher Boorse (Boorse, 2014). This conception allows for the precision of limits and numerical determination of 'pathology'—disease, without the subjective and imprecise considerations of 'illness' (Gaudenzi, 2014). Everything now can (must) be read, translated, or reduced to data. Even issues related to mental and psychic processes, which were excluded from Boorse's scope, are now included. We are now dealing with the extraction, processing, flow, and storage of data (Vianna Sobrinho, 2021).

This situation is comparable to the last decades of the 20th century, where the technological boom in electronics, modern chemistry, biology, and genetics, leading to practical applications and research in health, brought a chain of moral conflicts and dilemmas, as well as unforeseen harmful side effects on life and strong impacts on social relations. Within a generation, powerful technologies and new scientific practices no longer fit the moral codes of previous centuries. This sociotechnical disarray is at the root of the debates that marked the foundation of the field of bioethics in the 1960s and 1970s.

At the beginning of the next decade, philosopher Stephen Toulmin made an interesting analysis of the transformations of this period in an article arguing that "medicine saved the life of ethics" (Toulmin, 1982). According to him, in academic debate until that period, "the central concerns of philosophers had become so abstract and general [...] that they had, in fact, lost all contact with the concrete and particular issues that arose in current practice, whether in medicine or elsewhere" (Idem, p. 749). The development of technologies that transformed our practices in the health field, in addition to expanding the real possibilities of intervening in the limits of life, also pointed to unprecedented issues in the field of research. Thus, this author assumes that "medicine 'saved the life of ethics' and returned to ethics its seriousness and human relevance [...] By reintroducing controversial topics raised by particular cases into ethical debate, forcing philosophers to return to the Aristotelian problems of practical reasoning" (Idem, pp. 749–750).

Given all this, it seems undeniable that philosophical reflection has once again become necessary, but with greater breadth, aided by epistemology and ontology to underpin ethical conduct. The main point to which we must pay close attention is precisely the boundary that separates us from any other existence. The relationships between humans and their technologies have never reached such extreme points before, where what characterizes us, cognitive capacity, is presented in technical objects that "become independent because they are so dangerously similar to us" (Malabou, 2019).

Building Human Rights

To develop a proposal that aligns under the banner of defending Human Rights (HR), it is essential to reclaim the power of a justice claim still considered and worked on within the moral field. This task is almost altruistic in itself when assumed as a task of bioethics. Agreeing with the almost hegemonic sociological conception that HR are socially constructed and are not the same as legal rights (Nash, 2015), we highlight the need to seek evidence for our argument that we are living in a new historical moment. It is also necessary to remember the continuous nature that has characterized the history of HR defense, where construction always occurs in this historical process and depends on active movements of social struggles and achievements. Just as the Universal Declaration of 1948 "crystallized 150 years of struggle for rights", it should be considered "more the beginning of the process than its apex" (Hunt, 2009). We tend to accept that a new conception of rights for humanity will begin precisely when we become aware of the necessary demarcation of the ontological boundary that guarantees these rights.

Thus, by the very delineation of the exposed problem, the first significant evidence we call attention to is the universal character of the issue we face at this moment in history. We know that this universal character is recent and of utmost importance (Sen, 2001), and it may never have been so indisputably evident since the desires of Kantian cosmopolitan liberty (Bobbio, 2004). This starting point presents itself as the task to be defended with the premise that we have reached a moment of historical evolution (to avoid misdirection in the debate around the concept of 'progress'). We are no longer defending Western-tinted forms. Indeed, it is not a Westernizing, colonialist, liberal, bourgeois, or individualistic position. Understanding the ontological demarcation where the defence of this right is located would practically remove us from criticisms or the mistake of framing disputes at the level of social or multicultural class interests. We are explicitly postulating rights that protect and delimit the entirety of human capabilities, in contrast to something that threatens the very *ethos*.

The reclaiming of the argument for universality as a foundation gains strength through its utopian character, advocating a globalization conceived as counter-hegemonic in its own hue: with the cosmopolitanism of humanity as a common heritage for all (Santos, 2000, p. 305). From this perspective, we might even find the space to embrace a more radical critique that precedes the conflictual relations between the individualistic interests of certain human rights—structuring capitalism itself—and the peripheral rights, the political, social, and collective ones, considered more incidental (Mascaro, 2017, p. 126).

The scope of this utopian position must have a programmatic character, grounded in realism, descending to the realm of programmatic power disputes. It cannot rely solely on the power of thought to transform ideologies and reality. This call for a more 'scientific' utopia, working in the realm of political realism, must recognize the need for mobilizing international human rights movements to radicalize their agenda. It must go beyond the individualistic framework of privileging political and civil liberties. Beyond personal well-being, the primacy of social solidarity will only come from the struggle for global economic well-being (Moyn, 2014).

This form of cosmopolitanism, rooted in the fight for humanity's common heritage, can intensify interregional interactions. It can surpass globalization models that perpetuate colonial conditions in North-South relations, overcoming the characteristic vices of globalized localisms and localized globalisms (Santos, 1997). It is precisely against the advance of colonialism that we propose the new struggle for human rights. Colonialism in the 21st century is restructuring the dynamics of capitalism through the evolution of socio-economic dominance processes, with a culture of total life management through datafication (van Dijck, 2014). Beyond what we witnessed during the industrial period, where human labor was transformed into an abstract social dimension through commodification, digital age capitalism can be conceived as "data colonialism, which is transforming human nature (i.e., pre-existing currents of human life in all its diversity) into a newly abstract social form (data) that is also ripe for commodification" (Couldry & Mejias, 2019, p. 32).

We observe that current efforts by various civil society entities, diverse scientific associations, and state and supranational organizations are working on creating regulatory mechanisms and satisfactory demarcations for the legal use of AI technologies. There is a vibrant movement to construct regulations that try to keep pace with the rapid discoveries of the last decade. This new culture increasingly exposes conflicts and weaknesses in both individual and collective rights protections. However, these efforts often seek to incorporate the already established production of these technologies into the legal framework at the final data control stage. This results in prioritizing the management of distribution, a process that inherently carries the inequities of this mode of production. With the progression of automation in the capital production process, without ensuring that the ownership of this production serves collective social interests, this approach will inevitably lead to the radicalization of this economic model of concentration. It confronts us with the possibilities of an autonomous technology that promises to surpass and render humanity obsolete (Dyer-Witheford, 2019).

Final Considerations

The starting point must be a new beginning. We need to return to the fundamentals to justify our intention to protect all of humanity, seeking a new understanding for new sentiments (Hunt, 2009, p. 19). Applying rights that keep us "equal when differences demean us, and different when equality detracts from our identity" (Santos, 1997), it is insufficient to adapt regulations to confront AI technologies, accepting them at the level of existing rights, listed from the first to the fourth generation. We advocate for developing a body of validated knowledge and composing a proposal with a minimum and essential set of new Human Rights. These should be considered of higher intensity, even preceding first-generation rights.

Starting from a renewed list of considerations and 'considerands' that demonstrate the demarcation of the ontological boundary as the great current challenge, we would mostly address negative rights. The primary concern is the definition and guarantee of our nature in the face of the power of the technology we have created. We would thus take on the mission of reclaiming a primordial claim that guarantees the *ethos* of humanity before we can no longer define it in its intrinsic dignity. This would be a definitive advancement over the historical debt of establishing universal values (Gabriel, 2022) without dependence on social historical arrangements or anthropological observation but built on the legitimate desire to qualify humanity as inherently valuable.

The right to be limited in capacities, imperfect in behaviors, insecure in relationships, to have needs, and to depend on the collective—the simple and non-transferable right to belong to humanity. We must never be compared to or confronted by the power of the technologies we create. We cannot give space to the character we construct for our existence. The limits of our possibilities must demarcate the very limit of the right to be human.

References

Arendt, H. (2004). *A Condição Humana*. Rio de Janeiro: Forense.

Bobbio, N. (2004). *A Era dos Direitos*. Rio de Janeiro: Elservier.

Boorse, C. (2014). A second rebuttal on health. *J. Med. Philosophy*, 39(6), 683–724.

Bostrom, N. (2004). The future of human evolution. *In*: Tandy, C., Bostrom, N. (Eds.), *Death and Anti-death: Two Hundred Years after Kant, Fifty Years after Turing* (pp. 339–371). Palo Alto: Ria University Press.

Camargos, E.F. (2017). A doença tem indicação de UTI, mas o paciente não..., caro colega. *Geriatr. Gerontol. Aging*, 11(4), 156–157.

Chaplin, C. https://www.charliechaplin.com/en/articles/29-the-final-speech-from-the-great-dictator-

Coeckelbergh, M. (2020). *AI Ethics*. Cambridge, MA: The MIT Press.

Couldry, N., & Mejias, U.A. (2019). *The Costs of Connection. How Data is Colonizing Human Life and Appropriating It for Capitalism*. California: Stanford University Press.

Cyranoski, D., & Reardon, S. (2015). Chinese scientists genetically modify human embryos. *Nature*, 22 abr. 2015. Disponível em: https://www.scientificame-rican.com/article/chinese-scientists-genetically-modify-human-embryos/. Acesso em: 23 abr. 2019.

Dyer-Witheford, N., Kjosen, A.M., & Steinhoff, J. (2019). *Inhuman Power: Artificial Intelligence and the Future of Capitalism*. London: Pluto Press.

Floridi, L. (2011). *The Philosophy of Information*. Oxford University Press.

Floridi, L. (2013). *The Ethics of Information*. Oxford University Press.

Floridi, L. (2014). *The Fourth Revolution: How the Infosphere is Reshaping Human Reality*. Oxford University Press.

Francisco. (2024). *Artificial Intelligence and Peace. Message of His Holiness Pope for the 57th World Day of Peace*. Vatican. Acesso: LVII World Day of Peace 2024: Artificial Intelligence and Peace | Francis (vatican.va) em 01/01/2024.

Gabriel, M. (2018). *Der Sinn des Denkens*. Berlin: Ullstein Verlag.
Gabriel. M. (2022). *Moral Progress in Dark Times: Universal Values for the 21st Century* (1st Edn.). Kindle Edition.
Gaudenzi, P. (2014). A tensão naturalismo/normativismo no campo da definição da doença. *Rev. Latinoam. Psicopat. Fund.*, 17(4), 911–924.
Hunt, L. (2009). *A invenção dos direitos humanos. Uma história*. Trad. Eichenberg, R. São Paulo: Companhia das Letras.
MacIntyre, A. (1976). *Historia de la* ética. Barcelona: Paidós.
Malabou, C. (2019). *Morphing Intelligence. From IQ Measurement to Artificial Brains*. New York: Columbia University Press.
Mascaro, (2017). Direitos humanos: Uma crítica marxista. *Lua Nova: Revista de Cultura e Política*, São Paulo, 101, 109–137.
Moyn, S. (2014). The future of human rights. *SUR. International Journal on Human Rights*. 20, 57–64.
Mukherjee, S. (2018). *O gene, uma história* íntima. São Paulo: Cia das Letras.
Nash, K. (2015). *The Political Sociology of Human Rights*. United Kingdom: Cambridge University Press.
Rushkoff, D. (2018). Os ultrarricos preparam um mundo pós-humano. *OutrasPalavras*, 19 nov. Disponível em: https://outraspalavras.net/sem-cate-goria/os-ultra-ricos-preparam-um-mundo-pos-humano/. Acesso: 18 mar. 2019.
Sagan, C. (1996). *Pálido Ponto Azul: Uma visão do futuro da humanidade no espaço*. São Paulo: Companhia das Letras.
Sales Júnior, J.A.L. (2017). *Reflexões bioéticas sobre o morrer na vigência do paradigma biotecnicocientífico nas UTIs*. Tese (Doutorado em Bioética Aplicada e Saúde Coletiva) – PPGBIOS, Fundação Oswaldo Cruz. Escola Nacional de Saúde Pública Sergio Arouca, Rio de Janeiro.
Santos, B.S. (1997). Uma concepção multicultural de direitos humanos. *Lua Nova: Revista de Cultura e Política*, 39, 105–124. https://doi.org/10.1590/S0102-64451997000100007.
Santos, B.S. (2000). *A crítica da razão indolente. Contra o desperdício da experiência*. Porto: Afrontamento.
Savulescu, J. (2001). Procreative beneficence: Why we should select the best children? *Bioethics*, 5(6), 413–426.
Sem, A. (2001). Direitos humanos e diferenças culturais. *In*: Darnton, R., Duhamel, O. (Eds.), *Democracia*. Rio de Janeiro: Record.
Stiegler, B. (2014). *Symbolic Misery* (Volume 1), *The Hyper-industrial Epoch*. Cambridge: Polity Press.
Stiegler, B. (2019). *The Age of Disruption: Technology and Madness in Computational Capitalism*. Cambridge: Polity Press.
Topol, E. (2013). *Creative Destruction of Medicine: How the Digital Revolution will Create Better Healthcare*. New York: Basic Books.
Topol, E. (2019). *The Topol Review: Preparing the Healthcare Workforce to Deliver the Digital Future*. Disponível em: https://topol.hee.nhs.uk/wp-content/uploads/HEE-TopolReview-2019.pdf. Acesso em: 18/jun/2019.
Toulmin, S. (1982). How medicine saved the life of ethics. *Perspectives in Biology and Medicine*, 25(4), 736–750. Chicago.
van Dijck, J. (2014). Datafication, dataism, and dataveillance: Big Data between scientific paradigm and ideology. *Surveillance & Society*, 12(2), 197–208.

Vianna Sobrinho, L. (2019). A medicina de dados. O ocaso da clínica. *In*: Castro, J.C. & Guimarães, M.N. (Eds.), *Caminhos da Bioética* (Vol. 2). *Coleção FESO* (pp. 118–151) Teresópolis: Editora Unifeso.

Vianna Sobrinho, L. (2021). *A Medicina de Dados. O Ocaso da Clínica*. São Paulo: Zagodoni.

Vilaça, M.M., & Palma, A. (2012). Limites biológicos, biotecnociência e transumanismo: Uma revolução em Saúde Pública? *Interface – Comunic., Saude, Educ.*, 16(43), 1025–1038.

5 | Regulation for Data Protection and Privacy (In Health)

Priscilla Mesquita Buzzetti Matsushita*[*] and Cristyny L. Goncalves de Almeida

This chapter addresses the regulation of personal data protection, especially in the health sector. We intend to compare the specific legislations of Brazil (LGPD), Europe (GDPR), United States (HIPAA), and China (PIPL) with their implications for health data processing. The four legislations aim to protect personal data, each with distinct approaches. Despite the nuances, the laws share a focus on privacy, security, and transparency in information processing. The goal is to provide a detailed analysis of the differences and similarities between the legislations, highlighting the main aspects. It emphasizes the importance of observing data protection laws and regulations, as well as ethical principles in the health environment, to ensure data self-determination and avoid algorithmic discrimination.

Big Brother is watching you.
— George Orwell

Introduction

People have always sought privacy. On December 15, 1890, Warren and Brandeis published an article in the Harvard Law Review (USA) titled "The Right to Privacy". In this work, the authors defined privacy as the "right to be let alone", highlighting its importance in protecting individual rights (Warren & Brandeis, 1890).

It is surprising that the article was written in 1890, in reference to the invention of instant photography (which alone is no longer a concern), but technology continues to be a source of concern.

The right to privacy can be considered as the right of an individual to determine what information about them can be communicated to others (Schoeman, 1984).

*Corresponding author: priscillamatsushita@gmail.com

Regulatory evolution revolves around the concept of privacy, with the UN General Assembly's Declaration of Human Rights (1948) standing out. This declaration recognizes values of individual and family privacy protection and serves as the inspiration for all personal data protection laws in its Article 12.

In 1981, the Council of Europe adopted the Data Protection Convention (Convention 108), making the right to privacy a legal imperative[1].

Throughout history, the advancement of information technologies, increased connectivity, and the circulation of information have been closely linked to the right to privacy and data protection, causing concern among jurists. Health data is relevant to health insurance companies, life insurance, and even pharmaceutical industries.

Equally important, technology has introduced tools that allow qualitative improvements in management, such as the use of algorithms that rationalize, standardize, and enhance the quality of many work processes (Sarlet & Molinaro, 2019).

Furthermore, in information societies, operations, decisions, and choices previously left to humans are increasingly delegated to algorithms. These algorithms can advise or even decide how data should be interpreted and what actions should be taken as a consequence (Floridi et al., 2016).

As unprecedented events occur, such as the development of autonomous vehicles[2], it is necessary to reshape the legal system to support and enable the advancement of new technologies, thereby driving socioeconomic development (Benenson, 2008).

In this context of technological innovation, the importance of privacy and data protection has become even more evident, especially in critical and vulnerable sectors like healthcare. With the growing prominence of privacy and data protection issues, compliance with data protection laws and regulations is crucial within healthcare organizations. Patients' health data are extremely sensitive and confidential.

While the circulation of private information significantly benefits sectors, particularly the personal data industry, concerns about risks to individuals are growing. These concerns include the influence on self-determination, decision-making foundations, consumption habits, and social, political, and cultural understanding, as well as how users handle information (Garcel et al., 2020).

Numerous accusations have been made against the pharmaceutical industry regarding the use of health data to improve marketing campaigns or share data

---

[1] Authors' notes: Convention for the Protection of Individuals regarding Automatic Processing of Personal Data (Convention 108) of January 28, 1981.

[2] Authors' notes: In recent years, the field of robotics has seen significant growth. This progress is largely due to advancements in electronics, which have introduced new types of sensors, reduced the costs of embedded systems, and greatly enhanced available computing power. The boost in computing capabilities allows for processing larger amounts of data at faster rates and enables the formulation of optimization problems that were previously deemed unsolvable.

with health insurance providers. For example, a pharmacy's database can reveal if a consumer has chronic diseases, sexually transmitted infections, degenerative diseases, takes controlled medications, buys medication for sexual impotence, uses condoms or contraceptives, has hemorrhoid issues, or is depressive or suicidal. These pieces of information can potentially harm the data subjects to varying degrees.

Additionally, the use of algorithms and artificial intelligence in analyzing health data raises concerns about algorithmic discrimination. Algorithms can perpetuate existing biases, resulting in discrimination. Therefore, healthcare organizations must not only protect patient data privacy but also ensure their technologies are fair and equitable, avoiding discrimination and promoting justice and equality in healthcare access.

Compliance with data protection laws also aligns with bioethical principles, such as respecting patient autonomy. By safeguarding patient health data and adhering to sharing restrictions, healthcare organizations uphold patients' rights to control their medical information.

In much of the world, countries have privacy and data protection laws. These laws cover the collection and use of personal data. They aim to enhance protection and sensitivity for specific data categories, such as health data. Generally, these laws apply locally. As a rule, anyone actively conducting business in a country must follow its privacy and data protection laws (Densmore, 2019).

Protecting health data is essential. The appropriate use of digital health involves several dimensions: health promotion and disease prevention, security, ethics, interoperability, intellectual property, data security (confidentiality, integrity, and availability), privacy, cost-effectiveness, patient engagement, and accessibility (WHO, 2021).

Thus, we move on to analyze the landscape of health data protection legislation in Brazil, Europe, the United States, and China.

General Overview of Legislation

Brazil – LGPD (General Data Protection Law)

Law 13.709/2018, also known as the General Data Protection Law (LGPD), represents a significant milestone in the protection of personal data in Brazil. It aims to protect the fundamental rights of freedom, privacy, and the free development of an individual's personality (Brazil, 2018). The push for the creation of the LGPD in Brazil was largely driven by the growing global awareness of the importance of privacy and data protection.

The LGPD was highly inspired by European legislation, specifically the General Data Protection Regulation (GDPR), which came into force in 2018. The GDPR seeks to protect data subjects and promote the flow of information between countries with equal levels of protection (Rabaioli & Lopes, 2019).

In Brazil, discussions about personal data protection gained momentum in the early 2000s with the increasing digitalization of services and the popularization of the internet. Additionally, the need for specific legislation on data protection became evident with the rise in incidents related to privacy and information security, such as the global Cambridge Analytica case[3].

The law was inspired by the European GDPR but adapted to the peculiarities of the Brazilian context. Originally published in 2018, it came into force in 2020, due to the Covid-19 pandemic, to give companies more time to adapt to the new requirements. The LGPD establishes clear guidelines on the collection, use, storage, and sharing of personal data, giving individuals greater control over their personal information. It has a multi-sectoral application, meaning all companies, regardless of their size or field of operation, including public agencies and companies, must comply with the legislation.

Europe – GDPR (General Data Protection Regulation of the European Union)

The General Data Protection Regulation of the European Union (GDPR) set new standards for personal data protection and significantly impacted other jurisdictions, encouraging several countries, including Brazil and China, to update their privacy laws. Enacted as Regulation (EU 679/2016), the GDPR came into effect on May 25, 2018, replacing the 1995 Data Protection Directive (Directive 95/46/EC) (EU, 2016). This change reflected technological advancements and the growing importance of data privacy.

The Directive aimed to harmonize data protection laws across EU Member States, ensuring basic protection and facilitating free data movement. However, it allowed flexibility in implementation, resulting in significant variations between national laws. The technological landscape evolved dramatically with the rise of the internet, social networks, e-commerce, and massive data collection, rendering the 1995 Directive insufficient for new data processing realities.

High-profile data breaches and privacy scandals increased pressure for stricter regulation. To ensure uniform data protection rules across Europe, the GDPR emerged as one of the most comprehensive and influential data protection laws worldwide. Besides serving as a model for other regulations, many organizations

---

[3] Authors' note: In 2016, Cambridge Analytica was hired to help target political ads to potential voters (Rosenberg, Confessore, & Cadwalladr, 2018). The company used a methodology based on psychological profiles to create predictive models. These models allowed for the segmentation of voters based on their interests and behaviors. This process involved collecting personal data from Facebook users without their consent through "thisismydigitallife", a psychological quiz app that gathered information about users and their friends (Cadwalladr, 2018). Cambridge Analytica then used this data to create psychological profiles of users. They developed over two hundred and fifty algorithms that enabled the targeting of political ads based on the preferences and personality traits of potential voters (Hern, 2018).

outside Europe had to adjust their practices to comply with it when handling data from European citizens.

The GDPR proposal started in 2012 and was finally adopted in April 2016, with a two-year transition period for organizations to adapt to the new requirements. It came into force in May 2018. The GDPR introduced significant innovations for implementing cohesive public policies on personal data protection across the EU. Its implementation marked a milestone in privacy protection, reflecting the EU's commitment to defending individual rights.

United States – HIPAA (Health Insurance Portability and Accountability Act)

Concerns about health data protection in the U.S. began much earlier, in 1996, with the Health Insurance Portability and Accountability Act (HIPAA) (US, 1996).

In the United States, federal laws address privacy across various societal segments. HIPAA governs health data protection, along with the Health Information Technology for Economic and Clinical Health Act (HITECH). The HITECH encourages the adoption of Electronic Health Records (EHR) and the associated supporting technology.

HIPAA mandates that companies handling Protected Health Information (PHI) must implement and follow physical, network, and process security measures. It aims to protect individuals, usually patients of healthcare providers and members of health insurance plans or government health benefit programs, by establishing privacy and security rules.

HIPAA focuses on 'portability', allowing individuals to transfer their health insurance coverage from one employer to another without pre-existing health conditions hindering their professional transitions. Upon its enactment, HIPAA addressed various healthcare-related topics, including the advancement of standardized electronic transactions within the healthcare sector.

The central idea was that certain "standard transactions", such as submitting a health insurance claim and paying that claim, could be standardized into mandatory electronic formats, thereby creating efficiency savings and more effective outcomes. With these standardized transactions came concerns about the privacy and security associated with placing healthcare information in electronic format.

The most critical provisions of the law require healthcare providers to keep all personally identifiable medical information private and secure. According to the U.S. Department of Health and Human Services (HHS, 2020), HIPAA sets national standards for protecting individuals' medical records and other personal health information. It applies to health plans, healthcare clearinghouses, and healthcare providers conducting certain electronic health transactions.

In line with these regulations, HIPAA's technical safeguards require access control, ensuring only authorized personnel can access PHI. Access control includes using unique user IDs, emergency access procedures, automatic logoff, and encryption. Additionally, the requirements encompass audit reports or tracking logs that record activity on hardware and software.

Other technical policies for HIPAA compliance address integrity controls or measures implemented to confirm that PHI is neither altered nor corrupted. Incident recovery in IT and external backups are key components that ensure electronic media errors and failures are quickly corrected, allowing accurate recovery of patient health information.

Thus, HIPAA compliance is necessary not only for companies providing treatment, payment, and healthcare operations but also for partners and anyone accessing patient information and supporting treatment, payment, or operations. This means that subcontractors and other business associates must also comply with HIPAA regulations.

China – PIPL (Personal Information Protection Law)

The Personal Information Protection Law (PIPL) of China came into effect on November 1, 2021 (China, 2021), amid a global context of growing concern about privacy and data protection. It's important to note that the Chinese government, one of the largest digital economies in the world, had already implemented data security regulations, such as the Cybersecurity Law, in 2017.

The Chinese law was significantly influenced by international regulations, especially the GDPR, which also emphasizes the protection of personal data. The PIPL represents a substantial step forward in data protection in China, establishing a stringent legal framework for managing and securing personal data. Like other legislation, the Chinese law reflects both internal needs for security and privacy and adaptation to international data protection standards.

Data Protection in Healthcare

The issue of patient safety is emerging worldwide due to increased concern for respecting patient dignity, quality of care, and the stability of the healthcare system (Reis, 2013).

From a bioethical perspective, healthcare organizations have a duty to act in the best interest of patients (beneficence) and to do no harm (non-maleficence)[4]. Ethics and patient safety are closely intertwined. It is expected that all patients are treated with dignity (Kadivar, 2017) and protected from any possible harm. It is a moral obligation for healthcare professionals to prioritize doing no harm.

Beneficence and non-maleficence connect to patient rights, safe care, and health services that ensure care without danger or risk of injury, including the misuse or improper sharing of patient data.

---

4 Authors' notes: According to the principle of non-maleficence in medical ethics, ensuring patient safety and preventing any injury or harm is a major priority for healthcare providers (Kadivar, 2017).

In the United States, the *2024 HIPAA Journal Report*[5], citing data from the Department of Health's breach portal, highlights the number of violations that occurred and the volume of patient records exposed. Most violations resulted from employee negligence and non-compliance with HIPAA rules, as well as external hackers exploiting human vulnerabilities.

Additionally, the report notes that 11 incidents involved unauthorized access or disclosure of information, affecting a total of 36,533 patient records.

Compliance Requirements

In healthcare, compliance with data protection laws ensures the ethical and responsible use of patient health data, aligning with the principles of beneficence and non-maleficence.

Beyond protecting patient data, compliance improves trust between patients and healthcare institutions, enhances treatment transparency, and prevents abuses and discrimination arising from the use of sensitive data.

By adhering to data protection legislation compliance requirements, healthcare providers adopt better information security practices, contributing to the preservation of the information security triad: integrity, confidentiality, and data availability.

Case Reports

LGPD

Data subjects have various rights, including access to their data, correction of incomplete data, anonymization, blocking or deletion of unnecessary data, portability, information about data sharing, withdrawal of consent, and review of automated decisions. Similarly, the LGPD sets several requirements to ensure the protection of personal data. These requirements apply to all organizations processing personal data within Brazilian territory or involving the offering of products or services to individuals within the national territory.

Principles of the LGPD

The main compliance requirements of the LGPD include adherence to the following principles:

(a) **Purpose:** Personal data must be processed for specific, explicit, and informed purposes known to the data subject.

[5] Author's note: It is becoming increasingly difficult to ascertain the specifics of hacking incidents, as breach notifications often provide minimal details, such as whether ransomware or malware was involved. This lack of information makes it challenging for affected individuals to gauge their level of risk. Many breach notices describe these incidents as "cyberattacks that caused network disruption", implying they were likely ransomware attacks.

(b) **Adequacy:** Data processing must align with the purposes informed to the data subject.
(c) **Necessity:** Only necessary data should be processed.
(d) **Free Access:** Data subjects must have easy access to information about the processing of their data.
(e) **Data Quality:** Personal data must be accurate, clear, relevant, and up-to-date.
(f) **Transparency:** Data subjects must be informed about the essential aspects of data processing.
(g) **Security:** Technical and administrative measures must be adopted to protect personal data.
(h) **Prevention:** Measures should be taken to prevent damage.
(i) **Non-Discrimination:** Data processing must not be conducted for discriminatory, illegal, or abusive purposes.
(j) **Accountability:** Data processing agents must demonstrate the adoption of effective compliance measures.

In addition to the principles, the following legal bases for processing personal data must be observed: consent; compliance with a legal or regulatory obligation; execution of public policies; conducting studies by research bodies; execution or creation of contracts; regular exercise of rights; protection of life and health.

In case of security incidents that may cause significant risk or harm to data subjects, the organization must promptly notify the ANPD (National Data Protection Authority) and the affected data subjects.

When discussing security incidents, what is the first thing that comes to mind? Hacker attacks? A security incident can be defined as any confirmed adverse event related to a breach in personal data security. This includes unauthorized, accidental, or unlawful access that results in the destruction, loss, alteration, leakage, or any form of improper or illegal data processing, which can pose risks to the rights and freedoms of the data subject.

According to research conducted by Verizon, the majority of incidents are caused unintentionally by employees[6]. These incidents occur due to the lack of internal policies, inadequate training, improper data sharing, misuse of systems, unauthorized printing of documents, or even in situations where the employee falls victim to a sophisticated attack.

Thus, security measures must protect personal data against unauthorized access and accidental or unlawful situations of destruction, loss, alteration, communication, or any form of improper or illegal treatment.

Privacy by Design Concept

The LGPD emphasizes prevention by implementing technical measures to

---

[6] Author's notes: Since 2008, Verizon has conducted annual research on security incidents and publishes the findings in the Data Breach Investigation Report (DBIR). In 2022, the report indicated that 82% of incidents involved human error.

prevent damages due to personal data processing. This approach is based on Ann Cavoukian's Privacy by Design concept, which integrates privacy protection through the trilogy: information security systems, responsible business practices, and physical design and network infrastructure (Cavoukian, 2011).

To achieve its objectives, the concept includes seven fundamental principles designed to ensure that privacy is embedded in all aspects of system design and operation:

(a) **Proactive, not reactive; preventive, not corrective:** Prevent privacy events before they occur rather than reacting to incidents after they happen.
(b) **Privacy as the default setting:** Systems should automatically protect personal data without any action required from the individual.
(c) **Privacy embedded into design:** Privacy must be integral from the initial phase of system development.
(d) **Full functionality:** Privacy should coexist with other functionalities without compromising security or usability.
(e) **End-to-end security:** Full lifecycle protection: Ensure privacy from data collection to disposal.
(f) **Visibility and transparency:** Stakeholders must have clear visibility on how data is managed and protected. Business practices should be transparent and auditable, ensuring that privacy processes and policies comply with promises made to individuals.
(g) **Respect for user privacy – user-centric focus:** The needs and privacy of individuals must be at the center of processes and systems, ensuring their rights are respected and prioritized.

To comply with Brazilian legislation, it is essential to observe the principles for personal data processing and adopt the Privacy by Design concept as outlined above.

GDPR

The General Data Protection Regulation (GDPR) of the European Union establishes a set of strict requirements for the protection of personal data. The main compliance requirements of the GDPR include adherence to the principles of the legislation.

Only the minimally necessary data should be processed, within the limits of the specific intended purpose (principles of data minimization and purpose limitation). Data processing must occur legally, fairly, and transparently, and only for the period necessary to fulfill the purposes of the processing if it identifies the data subjects.

The data must also be accurate, complete, and processed with respect for confidentiality, and compliance must be demonstrated by the data controller. It is important to highlight that adherence to these principles provides greater clarity for those responsible for processing, enabling data subjects to make an informed decision about whether they agree with the respective processing.

In addition to adhering to the principles, the GDPR specifies six lawful bases for data processing:

1. Performance of a contract.
2. Compliance with a legal obligation.
3. Performance of a task carried out in the public interest.
4. Freely given, specific, informed, and unambiguous consent (explicit for sensitive data).
5. Legitimate interests.
6. Protection of vital interests of the data subject.

HIPAA

HIPAA (Health Insurance Portability and Accountability Act) establishes strict rules to protect the privacy and security of individually identifiable health information (Protected Health Information – PHI). It sets limits on who can access this information and under what circumstances it can be used. The law requires appropriate technical, administrative, and physical safeguards to protect personal health data. This includes the patient's right to their health information, the right to obtain a copy of their health records, request corrections, and protection against unauthorized access, misuse, and improper disclosure.

HIPAA consists of two main rules: The Privacy Rule and the Security Rule. The Privacy Rule sets standards for the use and disclosure of protected health information, while the Security Rule outlines requirements for the security of electronic health data. These requirements must be met not only by healthcare entities but also by third parties involved.

Non-compliance with HIPAA can result in significant penalties, including monetary fines and other sanctions, depending on the severity of the violation and the degree of negligence by the entity.

PIPL

The PIPL, like other legislations discussed here, establishes a strict regime for the protection of health data. It recognizes the sensitivity and potential harm if such data are compromised. Entities dealing with health data must adopt robust security and compliance practices, ensuring that the processing of these data always adheres to transparency, justification, and security.

The law defines legal bases for data processing, which include:

(a) Execution of a contract in which the individual is a party or implementation of human resource management according to labor rules and regulations established by law and in a collective contract made in accordance with the law.

(b) Other circumstances stipulated by laws and administrative regulations.

(c) Consent, which must be informed and explicit before processing personal data and can be revoked at any time.

Regarding principles, the Chinese law also mandates the application of fundamental data protection principles such as purpose limitation, data minimization, accuracy, transparency, and accountability. These principles are designed to ensure that personal data processing is conducted fairly, legally, and transparently.

Notably, the PIPL includes clauses allowing measures against countries that have acted in a discriminatory manner concerning personal information protection or that have harmed the rights and interests of Chinese data subjects and national security.

Common Points of Privacy and Data Protection Laws

Privacy and data protection laws across the globe share several common points, particularly in their primary goal of safeguarding the privacy and security of personal information, including health data. These laws aim to ensure that data is collected, stored, processed, and shared in a secure and ethical manner.

There are similarities in the inclusion of contractual requirements, audit protocols, and data subjects' rights. These rights include accessing personal information, correcting inaccurate data, requesting the deletion of unnecessary data, and revoking consent for data processing when applicable.

In the four jurisdictions studied, it is widely accepted that organizations are responsible for ensuring compliance with the standards and requirements set forth by the law. This responsibility includes implementing appropriate policies and procedures, appointing data protection officers, and conducting regular audits to ensure ongoing compliance.

Additionally, the requirements and principles for processing health data are similar. These include obtaining the data subject's consent, implementing adequate security measures to protect data from unauthorized access, limiting data use to specific purposes, and collecting only the data necessary for those purposes.

The concept of Privacy by Design is not unique to Brazil's General Data Protection Law (LGPD). It is also incorporated and recognized in the European Union's GDPR and aligns with the privacy principles of the HIPAA in the United States and the PIPL in China, albeit in a less explicit manner.

In the healthcare sector, data breaches can have severe consequences for both patients and healthcare entities. The legislations have introduced stringent financial penalties for violations, encouraging organizations to adopt proactive data protection measures.

HIPAA also imposes significant fines but places special emphasis on ensuring that healthcare entities implement robust security practices and respond effectively to any breaches of Protected Health Information (PHI).

Differences between LGPD, GDPR, and HIPAA

Although the American, Brazilian, and European legislations aim to protect the privacy and security of personal information, including health data, and establish guidelines to achieve these objectives, their primary focus remains on individual protection. In contrast, the Chinese legislation focuses on national security rather than individual rights.

A significant difference among the studied legislations is that while the HIPAA in the United States is more specific to the healthcare sector and has been in place for a longer time, the LGPD, GDPR, and PIPL have a broader approach. They apply to all sectors and types of personal data and were enacted more recently: 2018, 2016, and 2021, respectively.

HIPAA is applicable within the United States. However, if a foreign provider partners with a covered entity in the US or handles PHI transferred back to the US, they may be indirectly impacted by HIPAA compliance obligations. The American law has served as an inspiration for hospitals worldwide, and compliance with HIPAA demonstrates a higher maturity level in information security for institutions.

Regarding geographical application, there is a significant difference between the legislations. GDPR applies to all member countries of the EU and also to organizations outside the EU that process data of EU individuals. On the other hand, LGPD applies to all organizations processing personal data in Brazil, regardless of their geographical location. This means that both Brazilian and foreign companies dealing with personal data of Brazilian individuals must comply with the provisions of LGPD.

PIPL is distinguished from foreign laws in its unique considerations regarding national security, especially concerning the localization and international transfer of personal data[7].

The definition of sensitive data also varies among the regulations. The GDPR specifies categories such as race, ethnic origin, political opinions, religious beliefs, health, sexual orientation, among others. The LGPD expands on this by also including information related to biometrics and genetics as sensitive data. For the PIPL, financial account information is also considered sensitive.

---

[7] Authors' notes: The legislator chose to dedicate an entire chapter to the topic, specifically Chapter III, titled "Rules on the Cross-Border Provision of Personal Information". Furthermore, Chapter IV addresses entities that handle the personal data of Chinese citizens outside China's borders. The relevant provision states: "Article 53: Personal information handlers outside the borders of the People's Republic of China, as provided in Article 3, Paragraph 2, of this Law, shall establish a dedicated entity or appoint a representative within the borders of the People's Republic of China to be responsible for matters related to the personal information they handle, and are to report the name of the relevant entity or the personal name of the representative and contact method, etc., to the departments fulfilling personal information protection duties and responsibilities" (DigiChina, 2021). For more information, please refer to DigiChina, available at: DigiChina.

The GDPR includes the right to erasure, or the right to be forgotten, allowing individuals to request the deletion of their personal data under certain circumstances. This explicit provision is not present in the other legislations.

Regarding data breaches, it is common among the regulations to require breach notifications. All legislations provide for penalties. Under the LGPD, administrative fines can reach up to 2% of a company's revenue in Brazil, limited to R$ 50 million per infraction (ART. 52, II LGPD). Additional administrative sanctions include suspension of database usage, blocking access, issuing warnings, publicizing the infraction, permanent deletion of the database, and prohibition of data processing activities.

The GDPR imposes fines that can go up to 20 million euros or 4% of the company's global annual revenue, whichever is higher, for the most severe infractions.

Under U.S. law, penalties vary based on the level of negligence and are categorized into four levels, ranging from $100 to $1.5 million annually. Besides financial penalties, entities may suffer reputational damage, legal costs, and corrective action agreements.

In the PIPL, if an entity violates the requirements, regulators can impose corrective actions, including warnings, confiscation of illegal income, suspension of services, or fines up to 50 million RMB (Renminbi) or 5% of the previous financial year's revenue (Article 66).

Most legislations regulate international data transfers with the same goal: ensuring data security during transfers between countries. They require that all necessary security and privacy measures be maintained. Additionally, other requirements include adhering to local privacy laws to ensure compliance with both jurisdictions and conducting risk assessments, for example.

These legislations impose obligations on organizations handling personal data. The differences lie in the details, reflecting the distinct approaches and legal contexts of each region.

Data Protection and Artificial Intelligence in Healthcare

Artificial intelligence (AI) is seen as a major promise for the future, especially in healthcare, due to its ability and speed in processing data through algorithms. In the healthcare sector, the rapid evolution of AI technology offers numerous opportunities, such as more accurate diagnoses, personalized treatments, and improvements in medical data management. However, regulation often does not keep pace with this innovation, resulting in regulatory gaps that leave important issues inadequately addressed, such as patient data security and privacy.

Regulations are essential to protect patients and ensure the quality of healthcare services. However, it is crucial to find a balance between promoting innovation and ensuring safety so that the adoption of new technologies can benefit patients.

Thus, a relevant topic to be closely monitored is the regulation of artificial intelligence, considering the great importance of observing ethics and fundamental values (such as the responsibility of developers, informational self-determination, and individual freedoms) for the development of technology and AI.

However, technology evolves at an extremely rapid and exponential rate, with dramatic peaks of innovation in increasingly shorter periods, which also transforms the regulatory and legislative landscape that always follows innovation belatedly (Palhares, 2021).

There is a common saying that "data is the new oil". This analogy arises because AI always needs to be trained with data. Consequently, if there is a fierce competition among global powers for AI dominance, it also manifests in efforts for massive and granular control over data. This directly impacts individual and collective privacy, necessitating a discussion on the role of the state in regulating and protecting citizens (Oliveira, 2023).

According to Professor Fábio Borges Oliveira (2023), "data should be encrypted and transmitted over decentralized and federated networks, even if it takes more time, because it is safer and preserves people's privacy".

The Information Security Handbook by Cert (2011) states that encryption is essential to ensure information confidentiality. The term encryption comes from Greek, meaning "to write secretly". Encryption involves writing text in codes or ciphers, allowing only authorized individuals to read it. This method ensures that messages transmitted in public environments cannot be understood by third parties.

Encryption is fundamental for protecting health data in the context of AI. It ensures that health information is encoded, making it unreadable to unauthorized persons. This is crucial when these data are used to train AI models, as it guarantees patient privacy and confidentiality. Furthermore, encryption helps comply with regulations like HIPAA, which mandate rigorous protection of health data. In summary, encryption safeguards health data against unauthorized access, preserving security and privacy in the use of AI.

Additionally, regarding security measures, data must be protected against malicious intentions through programs that utilize Artificial Neural Networks for protection. On this topic, Professor Fabio Borges de Oliveira adds insightfully, "Instead of sharing data, share the learning of neural networks. The training of neural networks should also be encrypted from start to finish, even if at a higher cost."

This innovative approach can enhance data security and improve privacy in the use of AI in healthcare.

Final Considerations

This chapter highlights the importance of data protection laws in balancing privacy and economic development on the global stage. The comparative analysis of the LGPD, the GDPR, the HIPAA, and China's PIPL reveals both the convergences and particularities of each legislation in the context of data protection in the healthcare sector.

We discussed the emergence of these four significant laws and the driving forces behind them. The GDPR, implemented by the EU in 2018, was a comprehensive response to increasing digitalization and growing concerns about personal data privacy. The LGPD, in effect since 2020, was inspired by the GDPR, aiming to align Brazil with international best practices in data protection. HIPAA, enacted in the United States in 1996, initially focused on health insurance portability but later expanded to include the privacy and security of electronic health data. Finally, the PIPL established China's role as one of the strictest countries in personal data protection.

Regarding compliance requirements, despite contextual differences, the LGPD, GDPR, HIPAA, and PIPL share the common goal of protecting personal data and ensuring individual privacy. These laws mandate organizations to implement robust data security measures, uphold individuals' rights to their data, and maintain transparency in data processing practices. This convergence underscores a global commitment to safeguarding personal information in the rapidly evolving digital age, particularly within the sensitive realm of healthcare.

However, compliance requirements vary among the legislations. The GDPR and LGPD adopt a broader approach, covering all types of personal data, with an emphasis on principles such as transparency, purpose, and consent. Both laws impose severe penalties for violations. HIPAA, on the other hand, specifically focuses on the healthcare sector and establishes detailed standards for the protection of PHI. Its compliance requirements include implementing administrative, physical, and technical safeguards, as well as specific policies and procedures for risk management and security incident response.

While PIPL shares these principles, it places significant emphasis on national sovereignty. Harmonizing data protection laws is crucial in an increasingly globalized and digitalized world. Despite arising in different contexts and having unique features, these laws converge in their mission to protect individuals' privacy and ensure data security. In the healthcare sector, where data sensitivity is extreme, compliance with these laws is not just a legal requirement but an ethical responsibility.

Therefore, it is imperative for healthcare organizations to stay updated and continuously invest in their data protection policies, ensuring not only legal compliance but also the trust and security of their patients.

References

Alder, S. (2024, March). March 2024 healthcare data breach report. *The HIPAA Journal*. Retrieved May 24, 2024, from https://www.hipaajournal.com/march-2024-healthcare-data-breach-report/.

Autoridade Nacional de Proteção de Dados. (2022). *Semana de proteção de dados 2022: Incidentes de segurança da informação*. Retrieved May 17, 2024, from https://www.gov.br/anpd/pt-br/assuntos/semana-da-protecao-de-dados-2022/incidentes-de-seguranca-com-dados-pessoais#:~:text=Um%20incidente%20de%20seguran%C3%A7a%20com,dados%20inadequada%20ou%20il%C3%ADcita%2C%20os.

Benenson, R. (2008). *Perception pour véhicule urbain sans conducteur: Conception et implémentation* (Doctoral dissertation), Ecole des Mines de Paris, Paris Tech, Paris, France). Retrieved May 16, 2024, from https://www.researchgate.net/publication/47684562_Perception_pour_vehicule_urbain_sans_conducteur_conception_et_implementation.

Brazil. (2018). *Lei Geral de Proteção de Dados Pessoais* (LGPD), Lei No. 13.709, de 14 de Agosto de 2018. Retrieved from http://www.planalto.gov.br/ccivil_03/_ato2015-2018/2018/lei/L13709.htm.

Cadwalladr, C., & Graham-Harrison, E. (2018). Revealed: 50 million Facebook profiles harvested for Cambridge Analytica in major data breach. *The Guardian*. Retrieved May 15, 2024, from https://www.theguardian.com/news/2018/mar/17/cambridge-analytica-facebook-influence-us-election.

Cavoukian, A. (2011). Privacy by Design: The 7 foundational principles implementation and mapping of fair information practices. Information & Privacy Commissioner, Ontario, Canada. Retrieved May 20, 2024, from https://privacy.ucsc.edu/resources/privacy-by-design---foundational-principles.pdf.

CERT. (2011). *Cartilha de segurança da informação*. São Paulo: Comitê Gestor de Internet do Brasil. Retrieved May 21, 2024, from https://cartilha.cert.br/livro/cartilha-seguranca-internet.pdf.

China. (2021). *Personal Information Protection Law of the People's Republic of China* (adopted at the 30th session of the Standing Committee of the 13th National People's Congress on August 20, 2021, effective November 1, 2021). Retrieved from http://www.npc.gov.cn/englishnpc/c23934/202108/166e7b5c27b547e8bfb35ea76b86db02.shtml.

Convenção para a Proteção das Pessoas relativamente ao Tratamento Automatizado de Dados de Caráter Pessoal (Convenção 108) de 28 de janeiro de 1981. Retrieved May 19, 2024, from https://www.coe.int/en/web/conventions/full-list/-/conventions/treaty/108.

Creemers, R., & Webster, G. (2021, November). Translation: Personal information protection law of the People's Republic of China. Digichina. Retrieved May 24, 2024, from https://digichina.stanford.edu/people/rogier-creemers/?page=1&sort_order=desc&sort_by=iso_date.

Densmore, R. (2019). *Privacy Program Management Tools of Privacy within Your Organization* (2nd Edn.). IAPP (International Association of Privacy Professionals).

European Union. (2016). Regulation (EU) 2016/679 of the European Parliament and of the Council of 27 April 2016 on the protection of natural persons with regard to the processing of personal data and on the free movement of such data, and repealing Directive 95/46/EC (General Data Protection Regulation). *Official Journal of the European Union*, L119, 1–88. Retrieved from https://eur-lex.europa.eu/eli/reg/2016/679/oj.

Floridi, L., Mittelstadt, D., Allo, P., Taddeo, M., & Wachter, S. (2016). The ethics of algorithms: Mapping the debate. *Big Data and Society*, 3(2). https://doi.org/10.1177/2053951716679679.

Garcel, A., Moro, S.F., Souza Netto, J.L. de, & Hippert, K.P. (2020). Lei geral de proteção de dados: Diretrizes e implicações para uma sociedade pandêmica. *In*: V.C. de S.K.,

A.G., & J.L. de S. Netto (Eds.), *Coletâneas de artigos jurídicos: Em homenagem ao Professor José Laurindo de Souza Netto* (pp. 319–344). Curitiba: Clássica Editora.

Hern, A. (2018). Cambridge Analytica: How did it turn clicks into votes? *The Guardian*. Retrieved May 15, 2024, from https://www.theguardian.com/news/2018/may/06/cambridge-analytica-how-turn-clicks-into-votes-christopher-wylie.

Kadivar, M., Manookian, A., Asghari, F., Niknafs, N., Okazi, A., & Zarvani, A. (2017). Ethical and legal aspects of patient's safety: A clinical case report. *Journal of Medical Ethics and History of Medicine*, 10. Retrieved May 17, 2024, from https://www.ncbi.nlm.nih.gov/pmc/articles/PMC6150915/pdf/JMEHM-10-15.pdf.

Kessler, S.R., Pindek, S., Kleinman, G., Andel, S.A., & Spector, P.E. (2020). Information security climate and the assessment of information security risk among healthcare employees. *Health Informatics*. Retrieved May 18, 2024, from https://pubmed.ncbi.nlm.nih.gov/30866704/.

Oliveira, F.B. (2024). *Para pensar em proteção de dados no contexto da IA. Estratégia Latino-Americana de Inteligência Artificial* (*ELA-IA*). Retrieved May 18, 2024, from https://outraspalavras.net/tecnologiaemdisputa/para-pensar-a-protecao-de-dados-no-contexto-da-ia/.

Palhares, F. (2021). *Compliance digital e LGPD* (pp. 358–377). São Paulo: Editora Thomson Reuters Brasil.

Rabaioli, L., & Lopes, L.C. (2021). Os conceitos da Lei Geral de Proteção de Dados: Noções instrumentais sobre o tratamento de dados pessoais. *In*: F. Menke & R. de F. V. Dresch (Eds.), *Lei Geral de Proteção de Dados: Aspectos relevantes*. Indaiatuba: Editora Foco.

Reis, C., Martins, M., & Laguardia, J. (2013). Patient safety as a dimension of the quality of health care: A look at the literature. *Publimed*, 8(7), 2029–2036. Portuguese. doi: 10.1590/s1413-81232013000700018.

Rosenberg, M., Confessore, N., & Cadwalladr, C. (2018). How Trump consultants exploited the Facebook data of millions. *The New York Times*, New York, 17 Mar. 2018. Retrieved Oct, 01, 2024, from https://www.nytimes.com.

Sarlet, G.B.S., & Molinaro, C.A. (2019). Questões tecnológicas, éticas e normativas da proteção de dados pessoais na área da saúde em um contexto de Big Data. *Direitos Fundamentais e Justiça*, 13(41), 183–212. Retrieved May 14, 2024, from https://dfj.emnuvens.com.br/dfj/article/view/811.

Schoeman, F.D. (1984). Privacy: Philosophical dimensions of the literature. *In*: F.D. Schoeman (Ed.), *Philosophical Dimensions of Privacy: An Anthology* (pp. 1–33). Cambridge: Cambridge University Press.

United States. (1996). *Health Insurance Portability and Accountability Act of 1996*. Pub. L. No. 104–191, 110 Stat. 1936. Retrieved from https://www.govinfo.gov/content/pkg/PLAW-104publ191/pdf/PLAW-104publ191.pdf.

U.S. Department of Health and Human Services. (2020). Summary of HIPAA Privacy Rule. Washington, D.C.: U.S. Government Printing Office.

Verizon. (2022). *Data Breach Investigation Report*. Retrieved May 15, 2024, from https://www.verizon.com/business/resources/T605/reports/dbir/2022-data-breach-investigations-report-dbir.pdf.

Warren, S.D., & Brandeis, L.D. (1890). The right to privacy. *Harvard Law Review*, 4(5), 193–220. Retrieved May 14, 2024, from https://www.cs.cornell.edu/~shmat/courses/cs5436/warren-brandeis.pdf.

World Health Organization. (2021). *Global Strategy on Digital Health 2020–2025*. Geneva: World Health Organization. Retrieved May 14, 2024, from https://www.who.int/publications/i/item/9789240020924.

6 | On the Subject of a Theory of the Drone

Pedro Fraga Vianna

The combat drone is a cognitive project, with its fundamentals embedded in the mechanism that performs it. The strategic questions and expectations based on the intervention capability of this new technology in various reality fields are, in some way, a simulacrum of the philosophical doubt regarding the drone as an object. But what is this mode, and what determines it? This work examines the main perspectives on the political and theoretical issues concerning drones and seeks to understand the determining factors of these views, especially those inherent in the drone itself.

> *Die menschliche Vernunft hat das besondere Schicksal in einer Gattung ihrer Erkenntnisse: dass sie durch Fragen belästigt wird, die sie nicht abweisen kann; denn sie sind ihr durch die Natur der Vernunft selbst aufgegeben, die sie aber auch nicht beantworten kann; denn sie* übersteigen *alles Vermögen der menschlichen Vernunft*[1].
>
> — E. Kant

Drones are unmanned vehicles remotely piloted. The name is a neologism in English that expanded the word's meaning. 'Drone' is a metaphor combining metonymy and onomatopoeia: the male bee and its buzzing sound[2]. It's a

*Email: pfragavianna@gmail. com

1 Author's notes: "Human reason has the peculiar fate in one species of its cognitions that it is burdened with questions which it cannot dismiss, since they are given to it as problems by the nature of reason itself, but which it also cannot answer, since they transcend every capacity of human reason." Critique of Pure Reason (A viii) (Kant, 1998, p. 99).

2 Author's notes: The most significant element is the metaphor with the drone: The idea of the stingless creature, killed by the female, shaped the aeromodels used as targets in tests for main fighter jets (Chamayou, 2013, p. 24).

layman's term that lacks support in specialist terminology[3]. Perhaps it would be more accurate to call it a colloquial term. This term has been in use since the early development of the mechanism. It is widely known and referred to in discussions about the machine and has not disappeared from technical elocutions. It has become a recognized word among diverse audiences. The term has been transferred to many languages and their cultures as an Anglicism or translation. This phenomenon of linguistic historicity has much to be interpreted by anyone who wants to understand the drone itself.

This chapter focuses on combat drones, excluding other uses or types. However, it is inappropriate to define a strict restriction because it does not align with the object's characteristics. Tactical and strategic functions are intrinsic to drones in general. Similarly, the focus is not on one model or another, but on the concept of the drone. This entails considering its development diachronically, including its technical evolution and its proximity to other autonomous, robotic, or so-called Artificial Intelligence weapons.

The fascination with these technologies is notably complex. The allure of the unprecedented and the rapid metamorphosis of contemporary technologies contribute to the "drone mania" (Pultarova, 2014), though its precise measure is difficult to determine. This fascination also stems from a relative awareness of the significance of technological transformation. The drone itself operates within the expanding and contracting boundaries of the imaginable. Thinking about the drone is also about contemplating a fantasy and the ones it brings along.

The drone is a cognitive project, with its elements inscribed within the mechanism that performs it. Its technical components are epistemological, executing a program of orientation and intervention in reality, driven to revolutionize the way we think. The drone is claimed to have the advantage of being or becoming a form of representation to which we do not have access. A theory about the characteristics of the device—such as accuracy, efficiency, and potential—will perceive these elements as being reserved for a superhuman representational universe within the drone. The panoptic ideal of this technological generation aims to surpass other perspectives and pave the way for a monopoly on decision-making driven by Artificial Intelligence.

The judgments about drones are numerous and subject to such a variety of interests that it establishes an assumption of impertinence or irrelevance for philosophical speculation. The many and rival views are pressured by practical

---

[3] Author's notes: Let's pay attention to the intention behind defining a technical name or terminology. Specialists use terms like Unmanned Aerial Vehicle (UAV), Unmanned Combat Vehicle (UCV), Unmanned Combat Aerial Vehicle (UCAV), and preferably Remotely Piloted Aircraft (RPA). This preference arises because, to say the least, "unmanned could incorrectly imply that the system has no man-in-the-loop". Mike Fowler summarizes it this way: "Although popularly known as drones, the military refers to them as Remotely Piloted Aircraft (RPA) or Unmanned Aerial Vehicles (UAV). In military parlance, the term 'drone' is used for remotely controlled aircraft designed for target practice use" (Fowler, 2014).

demands, which take precedence and push theoretical, speculative interests to the background. However, the very conflict among practical judgments implies and invokes the idea of knowledge to compensate for their deficiencies and to support better arguments. Even though the purely cognitive value is muted and distant from the perspectives in conflict, all of them owe and are subordinate to the unknown of what should be the proper and accurate cognitive accompaniment of the issue.

There remains the difficulty of determining the perspective capable of doing so adequately. The first and foremost initiative to present a truly theoretical viewpoint on drones as a problem is *Théorie du drone* by the philosopher Gregoire Chamayou. The book specifically addresses UCAVs and their use in the military campaigns of the United States and its geopolitical allies, mainly Israel. Drones, operated from a geographically remote location relative to the target, bring about a historical change.

> (...) *pour qui fait usage d'une telle arme, il devient a priori impossible de mourir en tuant. La guerre, d'asymétrique qu'elle pouvait* être, *se fait absolument unilatérale. Ce qui pouvait encore se présenter comme un combat se convertit en simple campagned'abattage*[4] (Chamayou 2011, p. 24).

The artifice removes the attacker from the traditional theater of operations. The defender, who may not even perceive themselves as such, is less a belligerent subject and more a semi-conscious or unconscious prey, an object of pursuit. War transforms into a hunt. The state policy embodied in the drone has self-designated as "manhunting theoretical principles".

Chamayou absolutely condemns the supposed *nécroéthique*[5] of the drone. Opposing the image of a humanitarian weapon and a clean war, Chamayou reconsiders the illegality of killing someone in war when there is no risk of dying. He highlights the implications of this new condition for the warrior's *ethos*, suggesting that robots and remote control might become the coward's trump cards. He envisions new wars where only civilians are at risk and considered the dangerous elements, while the operators remain protected in a safe zone, despite being the only ones armed with high-tech missiles and automatons.

Chamayou draws on Menahem Yaari to attribute a profound meaning to this turning point in the historical process:

> *Un code militaire qui discrimine, dans les casoù des dangers sontimposés à des civils innocents, selon que ces civils sont les "nôtres" ou les "leurs" apparaît*

---

[4] Editors' notes: Freely translated as follows: "For whoever uses such a weapon, it becomes a priori impossible to die while killing. The war, which could have been asymmetric, becomes absolutely unilateral. What could still present itself as a battle turns into a mere slaughter campaign."

[5] Editors' notes: "*Nécroéthique*" can be translated into English as "necroethics".

> *comme on ne peut plus inquiétant dans un context oùs'affirment des attitudes ethnocentriques et xénophobes dans l'establishment israélien traditionnel. Nous sommes en train d'assister à une dérive de l'universalisme et de l'humanisme vers l'esprit de clocher et le tribalisme*[6] (Chamayou 2011, p. 189).

And *Théorie du drone* concludes its judgment on what is at stake in necroethics this way:

> *le projet n'est rien de moins qu'un dynamitage du droit des conflits armés tel qu'il avait été stabilisé dans la deuxième moitié du XXe siècle. Une éviscération des principes du droit international par un nationalisme de l'autopréservationvitale*[7] (Chamayou 2011, p. 24).

The horizon of critical analysis is, therefore, delineated by the notion of "us here against them there". The deepest purpose of the drone lies in this tribal characteristic. According to Chamayou, this implies a regime not only very different but fundamentally opposed to humanitarianism. The extraordinary technical development that drones represent and drive does not correspond to any moral equivalent. Worse, this technical advancement is projected and built precisely on a rudimentary principle.

The core divergence between Chamayou's position and the one presented in this article is as follows: Chamayou's *Théorie du drone* was conditioned by the historical experience he could reference, rather than being a theory in the strict sense. The weaknesses in Chamayou's thesis stem from the mistake of assuming historical knowledge to be rational knowledge—the error of mistaking *cognitio ex datis*[8] for *cognitio ex principiis*[9] (Kant, 1998, p. 693, A 837, B 865). Chamayou reduced his thoughts on armed flying drones to the events and historical processes he was aware of and then interpreted the meanings of the conjuncture he witnessed. *Théorie du drone* thus implies a theory about the relationships between Washington, its geopolitical allies, and its immediate elective enemies at the turn of the millennium. Different conclusions would emerge from a theorist considering UCAVs in different experiences.

---

6 Editors' notes: Freely translated as follows: "A military code that discriminates, in cases where dangers are imposed on innocent civilians, depending on whether these civilians are 'ours' or 'theirs' appears deeply troubling in a context where ethnocentric and xenophobic attitudes are asserted within the traditional Israeli establishment. We are witnessing a shift from universalism and humanism towards parochialism and tribalism" (Chamayou 2011, p. 189).

7 Editors' notes: Freely translated: "The project is nothing less than a demolition of the law of armed conflicts as it had been stabilized in the second half of the 20th century. A gutting of the principles of international law by a nationalism of vital self-preservation."

8 Editors' notes: *Cognitio ex datis* can be translated into English as "knowledge from data".

9 Editors' notes: *Cognitio ex principiis* can be translated into English as "knowledge from principles".

Thus, a different theory of the drone must necessarily emerge. First and foremost, it is inevitable to consider a conflict where the target of the remotely controlled weapon is simultaneously the operator of another drone. When thinking about the drone conceptually, it becomes essential to ask about the situation where the target can not only evade pursuit but also has the capability to strike back with another drone from their location. Drone warfare can indeed involve manhunting, such as when the United States uses drones against paramilitary groups in the hills and deserts of Afghanistan, who lack reciprocal artillery—as Chamayou himself notes. However, it is undoubtedly a different kind of war if the adversary is another military power, especially one that also employs hunter drones. In a drone vs. drone confrontation, we re-establish, in new terms, the premise of minimal symmetry and the offensive-defensive dynamics inherent in traditional warfare.

Commentators' imaginations often prioritize and are shaped by the idea that this potentially matched rival is the same as the recent past or the one seemingly poised to become the enemy soon. Jean-Baptiste Jeangène Vilmer concludes his review of *Théorie du drone* with a paragraph suggesting that, to contest Chamayou's interpretations, the capability or anticipated capability of adversaries, and which adversaries, should alter the significance of drones:

> *Lui qui se scandalise du slogan du* Reaper, '*Que les autres meurent*', *dont il fait un titre de chapitre (II.2), pense-t-il sincèrement que lorsque l'Iran, la Chine, la Russie ou le Pakistan auront des escadrons de drones armés, leur slogan à eux sera quelque chose comme* '*Mourons autantqu'eux*'?[10] (Vilmer, 2013).

Michael Boyle, in "The Costs and Consequences of Drone Warfare", highlighted that "at least 76 countries have acquired UAV technology, including Russia, China, Pakistan, and India". He added, at that time in 2013, that:

> "according to one estimate, global spending on drones is likely to be more than US$94 billion by 2021. (...) With demand growing, a number of states, including China and Israel, have begun the aggressive selling of drones, including attack drones, and Russia may also be moving into this market" (Boyle 2013, p. 22).

Boyle attempts a broader perspective and identifies five long-term consequences of the arms race involving drones. The first consequence is:

> "(...) it is now obvious that other states will use drones in ways that are inconsistent with US interests. (...) Many of the other states now capable

[10] Editors' notes: Freely translated: "Does he, who is outraged by the Reaper's slogan, 'Let others die,' which he uses as a chapter title (II.2), sincerely believe that when Iran, China, Russia, or Pakistan have squadrons of armed drones, their slogan will be something like 'Let's die as much as they do'?"

> of employing drones or near-equivalent technology—for example, the UK and Israel—are considered allies. But this situation is quickly changing as other leading geopolitical players, such as Russia and China, are beginning rapidly to develop and deploy drones for their own purposes. (...) Both Russia and China will pursue this technology and develop their own drone suppliers which will sell to the highest bidder, presumably with fewer export controls than those imposed by the US Congress. Once both governments have equivalent or near-equivalent levels of drone technology to the United States, they will be similarly tempted to use it for surveillance or attack in the way the US has done. (...)" (Boyle 2013, p. 23).

The article is outdated, but this may work to its advantage, as it highlighted trends that have either strengthened or come to fruition. A good example is its discussion on the rising prominence of China and Russia in drone warfare. Additionally, other parts of the text note the involvement of Hezbollah. However, Boyle's forecast loses value as a prediction because it assumes the continuity of current contenders, clearly defined and typified. By immediately projecting today's circumstances onto tomorrow, it not only undermines observations about the future but also about the present and its tendencies.

Matthew Crosston writes about drones in Italy, Pakistan, China, and South Africa, among other countries and regions. This less homogeneous and less aligned list, along with his critique titled "Pandora's Presumption", prompts the following question: "What happens when the drone playing field is not so heavily slanted toward the American side?" (Crosston, 2014). Therefore, the author expands and clarifies the horizon for a response. However, this effort remains vague and insufficient. There is a suggestion, but not a definitive answer.

The situation where a pursuer also becomes pursued must be considered in all possible scenarios. This context outlines a set of occurrences where any individual or entity can be in an active or passive position regarding drone operation. Thus, the possibility arises that the drone could ultimately fall into anyone's hands, accessible to all as either perpetrators or victims. This scenario fulfills the expectation of anarchy associated with the equality of means in force and violence.

If the relative strategic superiority of the drone offers an advantage that divides holders from non-holders, it adds the problem of immediate possession. The military, or whoever guards the weaponry, also has this opportunity.

David Hastings Dunn expresses his concern with "disembodied aerial warfare and the unarticulated threat" (Dunn, 2013). Drones have proven vulnerable to hacker actions and other stealthy countermeasures, which have already compromised even the most technologically advanced models held by some of the most powerful and well-funded armed forces. More directly, the concern relates to the proliferation of drones. Instead of being captured or repurposed, drones are starting to be produced by a wide range of interested parties, from the main competitors in the arms race to hobbyists experimenting with homemade fabrication. Even the simplest drones, compared to other common resources, already function as highly dangerous items[11].

Their technological principle involves relatively low costs and high economic and military returns. This remains true regardless of whether each piece was designed for war or even for violence. As Boyle summarizes:

> "One factor that is facilitating the spread of drones (particularly non-lethal drones) is their cost relative to other military purchases. The top-of-the line Predator or Reaper model costs approximately US$10.5 million each, compared to the US$150 million price tag of a single F-22 fighterjet" (Boyle 2013, p. 22).

The alarm over the mass proliferation of drones makes sense. Issues like the indiscriminate commercialization of drones to state and private entities, illegal reproduction or smuggling of industrial designs and products, and the so-called 'dronization' of everyday artifacts are new, ordinary problems. These problems may affect and potentially alter political hierarchies from the bottom to the top. On the other hand, it is essential not to ignore the entanglement in which these new realities are established. Few political entities can afford the technical, economic, and political costs of systematically operating an MQ-9 Reaper[12]. This includes the necessary tactical and logistical support to minimize resistance risks and ensure mission completion. Furthermore, less powerful and wealthy political forces tend to depend on technological hubs, which accentuates stratification.

In summary, drone operations involve a set of sophisticated technical elements from various specialized sources. Mediocre and cheap versions can be as or even more destructive than others, depending on the combination of capability and cost with the effective use of the weapon. However, this dynamic needs to be balanced (Horowitz et al., 2016; Gilli & Gilli, 2016; Benjamin, 2012). It is unclear whether the possession and control of drones will become—more or less—concentrated and centralized than they are today. Consequently, the impact on other weapons and political dominance through their use remains uncertain. However, it is important to note that while some relationships are indeterminable a priori, there are trends that events have made increasingly probable.

Chamayou highlighted a news piece that reveals a significant factor in historical transformation: The American armed forces received consulting services from the sports broadcasting network ESPN for training in data processing and storage (Chamayou, 2013, p. 61). The drone industry benefits from the

---

[11] Author's notes: Although it was not written with a focus on military models, it is interesting to read about the multiplication of possibilities for the use of drones in Choi-Fitzpatrick (2014).

[12] Editors' notes: The MQ-9 Reaper is an unmanned aerial vehicle (UAV) developed by General Atomics Aeronautical Systems primarily for the United States Air Force. It's a part of the Predator family of drones and is designed for long-endurance, high-altitude surveillance, and precision-strike missions.

technological advantage of certain corporations and the notable dependence of public power on private institutions. The example above is just one among many.

The drone drives both continuities and discontinuities in the history of private activity in warfare. Technical, productive, commercial, and logistical controls contribute to political control and power. Transnational business conglomerates rank among the top creators, manufacturers, suppliers, buyers, holders, and users of drone military technologies. Companies like DJI, Intel, General Atomics, Rosoboronexport, Uber, Boeing, Parrot, EADS, UPS, and Alphabet are key players in this field. The increasing importance of such organizations, with varying connections to the public sphere, alters the dynamics of these relationships.

How would drone theories consider them as weapons in a war, say, between Americans and Russians, or between the Organization of American States and Amazon, between the Catholic League and the largest communications empire, between multinational companies controlled by a human trafficker and the libertarian hacker militias of Unified Korea? Essentially, how would theories reflect on UCAVs in any imaginable conflict involving more symmetrical states or any future belligerents?

It is equally consequential to imagine the general effects of adopting drone technology by potential free agents not only outside of the state and regular conditions, like partisans, but also beyond the necessary condition of land—both in its symbolic, telluric character and its territorial, juridical-political nature (Schmitt, 1975). We find ourselves halfway between the present and the future and between experience and expectation when discussing globalized groups, virtually gathered and instituted, centered around interests that propose and execute violence as a means or an end. How close are we to seeing these groups armed with remote-controlled bombers?

The scope of this excursus may be sufficient to highlight the colossal aggregate of open crossroads that strategic decision-making seeks to systematize. How many and which futures can speculative thought legitimately expect?

Fundamentally, what emerges is a problematic triptych: (1) the near blindness of perspectives on the drone and its context, (2) the rare acknowledgment of this blindness, often concealed by the pretense of clarity, and (3) the complete erasure of these first two problems. What would it mean, after all, to have a vision on this matter, let alone a theoretical vision?

The colossal difficulties faced in a study for the theory of the drone are merely a miniature of the challenges that a truly theoretical viewpoint must overcome. The range of judgments by debaters only mimics the form that a theory should take, that is, the format that a viewpoint should take to fulfill the clauses of completeness and totality. This is why even the best academic contributions will disappoint the reader who hopes to fully grasp the domain of the entire problem of drones and their implications in historical processes.

Furthermore, every argument about drones exists in an unusual forum, where, besides the variety of deliberations on the subject, there is the thesis that the object itself embodies. The constitution and performance of drones' rest on theses that the device incorporates and executes. The drone, per se, is a set of theoretical propositions.

Even *Théorie du drone* did not overcome this essential insufficiency, but it stands out from its peers for linking academic viewpoints to those that take shape with the drone itself or through its mediation. The book's greatest merit was highlighting the connection between the scientific complex that creates the UCAV and the one that forms the academic discourse on the topic. The blind spots in both military strategy and academic investigation are also blind spots of the drone—those perspectives embedded in the artifact and enabled by it as an instrument for its operators. From inventive conception to the consequences generated by the conceived device, multiple representational flaws and blind spots collide with the central idea of precision. Chamayou opens the chapter "*Précisions*" in this way:

> *C'est très précis, c'est très limitè en termes de dommages collatéraux', déclarait à propos des drones armés l'ancien directeur de la CIA, León Panetta. L'argument traîne partout: le drone, du fait de sa 'précision', réduirait les 'dommages collatéraux' et permettrait un meilleur respect du principe de discrimination. Ce lieu commun est une fausse évidence, qui repose en réalité sur un véritable nid de confusions conceptuelles. Il convient de les pointer méthodiquement, je veux dire avec précision*[13] (Chamayou 2013, p. 197).

The last sentence of the passage can serve as an emblem of Chamayou's claimed position. On one hand, the reintroduction of the term and the emphasis indicate that the author perceives the "conceptual confusions" as subject to the same defect of imprecision to which drones are susceptible. This means he sees theoretical precision as the same quality of precision lacking in the device. On the other hand, the fact that the sentence expresses the pretension to offer the final precise analysis reveals Chamayou's misunderstanding of the root of the problem and the form of its vicissitudes. This diminishes and even casts doubt on the significance of the academic's incidental observation of some of these modifications.

Let us address the imprecisions of the drone. It is essential to note right away that they are so numerous and significant that 'imprecision' is more than a euphemism; it is a mischaracterization. The precision of the drone does not lack a minor margin, as it might seem. Instead, it consists of a dispersed set of partial representations, synthesized into a very poor scheme of objectivity. However, it

[13] Editors' note: Freely translated: "'It is very precise, it is very limited in terms of collateral damage,' said former CIA Director León Panetta about armed drones. The argument is everywhere: the drone, due to its 'precision', would reduce 'collateral damage' and allow for better adherence to the principle of discrimination. This common belief is a false obviousness, which actually rests on a real nest of conceptual confusions. It is necessary to point them out methodically, I mean precisely."

is preferable to maintain the terms precision and imprecision, with this caveat, to preserve dialogue with the existing bibliography and established terminology.

The idea of precision at stake is an overdetermination of various principles. These include the adherence to piloting protocols, the integrity of the engineering design, the exact manufacturing of the project, the normal and expected functioning of the equipment, the identification and distinction of the target, the accuracy of the shot, the proportionality of the effect, the prevention of collateral damage, the suitability of the employment situation, the tactical-strategic success, the achievement of the military objective, and the attainment of the political outcome anticipated by the decision. Thus, when these principles are presented merely as '*precision*', as a single term, their specificities are silenced. We must disentangle this conceptual confusion.

Leon Panetta's statement, cited by Chamayou, focuses on precision as the limitation of collateral damage. This example might seem randomly chosen or used to leverage a transcribed quote, but the principle of preventing collateral damage is indeed a priority. Being a characterization by a former CIA director, the way it is framed suggests that all other principles of precision contribute to this fundamental aim: "very precise and very limited in terms of collateral damage" conveys the notion that drones are confined or tend to confine their violence intention. The praise for the ability to contain undesirable effects implies the alignment of other precisions. These include those that contribute to this capability, such as the accuracy of the shot, the pilot's exceptional performance, the unequivocal recognition of the enemy, and the clear differentiation of the target. It also encompasses those that safeguard what exceeds the intent, such as the events unfolding as expected and the full achievement of the political goal, which become more precise if the perpetrators can choose their actions meticulously.

Considering certain parameters of military history and aviation, the reasoning "very precise and very limited in terms of collateral damage" seeks for the drone the quality of being surgical. As noted by Hugh Gusterson, "among other weapons, the Reaper carries the Small Smart Weapon, which is said to be capable of killing a person who is in one room while sparing people who are in the next room" (Gusterson 2016, p. 20).

On the horizon, the postulate emerges of a war that does not kill 'innocents'—to use a concept recurring in drone apologists' arguments. According to this line of reasoning, the drone would enable a precise strike, or at least much more precise than other models of aerial bombing.

Clearly, it is essential to note an obvious point upfront. The new destruction can be just as severe or even worse than past destruction, if that is the intention. One or more drones can replicate historical massacres or set a new terrible record. Vehicles and weapons from the distant or recent past can be replaced by UCVs, or adapted and 'dronized', and repeat or surpass those feats. 'Dronization' does not inherently eliminate or reduce risks to innocents. However, an argument like Panetta's suggests that the drone, unlike previous weapons, allows for this possibility.

That is not the case. Even if the best model can land a Hellfire missile between someone's eyes, it doesn't prevent the devastation within a radius of 15 or 20 meters. Precisely setting the explosion's focal point does not guarantee that the image is clear enough or that pilots and other controllers can accurately identify who or what is on screen (Gregory, 2011). A drone may seem like a scalpel compared to a previous bomber or a nuclear weapon, but that does not mean that the previous bomber and the nuclear weapon are the drone's substitutes. It also does not mean that commandos, snipers, guerrillas, and many other tactical elements cannot appear as scalpels compared to any drone.

Regarding the principle of target discrimination, the accuracy is evidently lacking. In debates about U.S. drone campaigns, higher and increasing numbers of non-combatant deaths are presented on one side, criticizing drone policy. On the other side, there are lower civilian casualty rates and more support for statements from the White House and government agencies. While judging the merits is crucial, the most characteristic aspect is the uncertainty of data about these deaths. Regardless of the accurate record of events, what must be defined is the structural inaccuracy of the information about drone warfare as we know it, with positive and negative political implications for different arguers.

However, an apology might assume that the drone is, even in the worst-case scenario, the best available resource (Bell, 2012). There are strong indications to the contrary, and certainly, there are more precise methods than missile strikes, even among more sordid and treacherous options. The accounting version of the argument does not hold up, nor is it necessary to waste paper ridiculing the fallacy of the lesser evil.

The praise for precision can be based on the potential or what the proper use of drones can offer. Let's consider this hypothesis again in "Practice makes Perfect?". The article's title, a rhetorical question, sets up the opinion that the correct use of technology and already verified improvements give the U.S. drone warfare project strong legal, technical, and strategic reasons to counter its critics. The authors attribute "a declining civilian death toll" to factors such as support in espionage, intelligence, and information, the data captured in the operation that killed Bin Laden, and, finally, greater care in CIA attacks (Plaw, Fricker, & Williams, 2011).

The argumentation undermines the thesis. The obvious problem of drones being subject to misuse persists, but we can discount this inconvenience, as well as the fact that the authors limited their observation to CIA operations in Pakistan. It is less easy to overlook that "the good drone" depends on those parallel elements, especially those that could replace the primary. It is odd, given the conclusions presented, that Plaw, Fricker, and Williams inadvertently hint at the possibility of other forms of warfare or that the act of war itself might be replaced by more precise political action. This is not the path the authors admit to us. They envision a single future, executed by the desired performance of the drone. They discuss a supposed practical preference for the drone, which may be encouraged.

The problem is that the logic of good use, instead of contradicting, confirms the accidental nature of precision. The misuse of drones and their supposed technical program calls into question not only the same plan but also the strategic and epistemological measures. Regardless of who uses this technology and how, currently, the drone embodies the intention of violence that does not require clear distinctions. Rather, it embodies the intention of violence that can be satisfied with a level and degree of clarity tinted by ethnocentric identifications, lowered moral standards due to conflict, confidence that this weapon is the right or even the best means for a particular action, belief in the lack of military alternatives, or sheer immorality.

But "Practice Makes Perfect?", as the title suggests, does not primarily focus on prudent or supposed good use but rather on the improvement and technical realization of precision in concept. The article does not speak of a distant, improbable goal but of a development in full swing. The authors also point to new accessories and technological components as factors in the alleged improvement of drone warfare. What is at stake is the promise of the technological concept, the expectation that the technical weaknesses hindering its optimal realization will be overcome, thus rendering the criticisms circumstantial.

Let's note that the theme of improvement plays an ambiguous rhetorical role. For example, Gusterson (2016) reveals his contact with Freeman Dyson and writes that the notable interlocutor, "for the founder of the company that makes the Predator and Reaper drones, is surprisingly skeptical of drone warfare" (Gusterson, 2016, p. ix). Despite Dyson's sincerity, it is common for those who endorse drone warfare to moderate their praise and highlight obstacles to avoid or counter accusations that the weapon is unjust or that its development leads to injustice.

What is most relevant to discussing the argument, however, is whether the improvement can fulfill the promise of the concept. Chamayou left the door open for the conjecture that the qualification of the cognitive project, such as the auxiliary resources that inform the machine, would confer the missing precision. Chamayou does not fully endorse this view but echoes the premise of improvement.

The perspectives uncovered call for another critique: at a minimum, colossal inaccuracies persist even if the drone, improved to its utmost potential according to the latency and demand of the concept, technically allows for precise killing or detaining. This could be something akin to a lethal or paralyzing projectile, perhaps virtual, with any range, in both trajectory and target. It's like being able to turn off or suspend a person's life with a touch, a voice command, or a bit, from any location. The glaring issue is the tendency for improvement to exacerbate the characteristic imprecision of a technology that disregards the randomness of decision and contingencies. It also aggravates secondary inaccuracies, such as the obscurity of what occurs, insecurity about the limits of violence and human relations, the intangibility of the war zone, or even whether there is a war.

The topic of improvements could also conclude with the question of the transmutation of the drone through technological combination and the development of its and other Autonomous Weapons Systems (AWS) (Leys, 2018; Wen, 2017). These evolutions transform the unmanned combat vehicle into something else. However, it is essential to highlight the increasing role of current drones in the production of knowledge, particularly regarding their autonomy and the complexity of the relationship between autonomous functions and human decision-making. The critical aspect of the 'unmanned' category lies in the inclination of UCVs to reduce human involvement, potentially regenerating or even rendering UCVs obsolete. Interestingly, the drone itself may already be on its way out.

Be that as it may, the uncertainties do not align with the momentum of military offensives or the presumptions in discourse. Like the drone itself, academic reflection remains imprecise, even under loose epistemological parameters. A rigorous critique, even at a preliminary stage, dismisses the existing theories of drones available to us.

The carelessness and superficiality of the considerations, particularly from those dedicated to the issue, lead to a drastic and unjustified reduction of uncertainty regarding the future. The urgency of knowledge for our daily conduct and our opinions on current events conditions this cognitive poverty. Yet, this condition is rarely acknowledged or considered, even by those with theoretical or philosophical pretensions. Discourses on drone warfare often serve as tools of deception, frequently amounting to self-deception.

References

Bell, D. (2012, January 27). *In Defense of Drones: A Historical Argument*. The New Republic.

Benjamin, M. (2012). *Drone Warfare: Killing by Remote Control*. Verso.

Boyle, M. (2013). The costs and consequences of drone warfare. *International Affairs*, 89(1), 1–29. https://doi.org/10.1111/1468-2346.12002.

Chamayou, G. (2013). *Théorie du drone*. La Fabrique.

Choi-Fitzpatrick, A. (2014). Drones for good: Technological innovations, social movements, and the state. *Journal of International Affairs*, 68(1), 19–36.

Crosston, M. (2014). Pandora's presumption: Drones and the problematic ethics of techno-war. *Journal of Strategic Security*, 7(4), 1–24. https://doi.org/10.5038/1944-0472.7.4.1.

Dunn, D. (2013). Drones: Disembodied aerial warfare and the unarticulated threat. *International Affairs*, 89(5), 1237–1246.

Fowler, M. (2014). The strategy of drone warfare. *Journal of Strategic Security*, 7(4), 108–119. https://doi.org/10.5038/1944-0472.7.4.12.

Gilli, A., & Gilli, M. (2016). The diffusion of drone warfare: Industrial, organizational, and infrastructural constraints. *Security Studies*, 25(1), 50–84. https://doi.org/10.1080/09636412.2016.1134189.

Gregory, D. (2011). From a view to a kill: Drones and late modern war. *Theory, Culture, and Society*, 28(7–8), 188–215. https://doi.org/10.1177/0263276411423027.
Gusterson, H. (2016). *Drone: Remote Control Warfare*. The MIT Press.
Horowitz, M., Kreps, S., & Fuhrmann, M. (2016). Separating fact from fiction in the debate over drone proliferation. *International Security*, 41(2), 7–42. https://doi.org/10.1162/ISEC_a_00257.
Kant, I. (1998). *Critique of Pure Reason*. Cambridge University Press.
Leys, N. (2018). Autonomous weapon systems and international crises. *Strategic Studies Quarterly*, 12(1), 48–73.
Plaw, A., Fricker, M.-S., & Williams, B.-G. (2011). Practice makes perfect? The changing civilian toll of CIA drone strikes in Pakistan. *Perspectives on Terrorism*, 5(5/6), 51–69.
Pultarova, T. (2014). Challenges of 'dronevolution'. *Engineering & Technology*, 9(12), 16–17.
Schmitt, C. (1975). *Theorie des Partisanen*. Duncker & Humblot.
Vilmer, J.-B. (2013, December 4). *Idéologie du drone. À propos de: Grégoire Chamayou, Théorie du drone, La Fabrique. La vie des idées*. Retrieved from https://laviedesidees.fr/Ideologie-du-drone.html (Accessed: January 20, 2024).
Wen, G. (2017). Unmanned aerial vehicles and the future of airpower: A technological perspective. *Pointer*, 43(2), 45–57.

7 | Essay on the Cybernetic Integration of Health Technologies: A Contribution from mHealth

Leandro Modolo

In this chapter, we argue that mobile Health (mHealth) represents a new condition of possibility for the integration of health technologies. This is due to the automation of the feedback cycle between individual and population actions, enabled by Artificial Intelligence (AI) techniques embedded in mHealth applications. In other words, by providing individual digital assistance to the citizen, mHealth functions as clinical technology and simultaneously co-produces with the user a set of population data that integrates health technology. We call this "cybernetic integration in health". This new condition for the principle of comprehensiveness in health is unprecedented. It allows the use value produced on a population health problem to participate in a customized way in individual clinical actions. These actions can then dynamically adapt to sanitary needs and responses.

> …this process does not end here, in this provisional coagulation, in this partial and determined form of apprehension and manipulation of reality—scientific research—but continues towards its open, uncontrolled destiny, where it longs to pulsate, even if transiently, to the rhythm of Life.
>
> — Mendes-Gonçalves, 1994, p. 84

Introduction

According to the World Health Organization (WHO) Global Observatory for eHealth, mobile health or mHealth is "medical and public health practice supported by mobile devices, such as mobile phones, patient monitoring devices, personal digital assistants (PDAs), and other wireless devices" (WHO, 2011, p. 6). The opportunities opened by mobile health range from promoting

*Email: modolole@hotmail.com

certain lifestyles and monitoring well-being to strengthening the doctor-patient relationship in the adherence and management of treatments of various complexities. This allows for real-time, 24/7 sharing of data, information, and communication between patients and healthcare professionals. Additionally, mHealth contributes to updating electronic medical records, providing services in resource-limited settings, monitoring post-surgery rehabilitation, managing clinical decision support, and intensive care. It also helps empower patients in self-monitoring and self-care (Saxena & Saxena, 2020).

mHealth has been applied to address a wide range of health issues, including diabetes, respiratory problems, cardiovascular conditions, dermatological issues, mental and psychological disorders, and sexual and fertility issues (Rivera-Romero et al., 2023).

Moreover, innovations in this sector are advancing rapidly. Wearable sensors, or even those implanted in the body, are increasingly used across a broad spectrum of health applications and are interconnected with various communication technologies. Examples include sensors for monitoring different physiological and biometric parameters such as cardiovascular metrics, body temperature, blood oxygen saturation, and many others (Deloitte, 2014). All this potential is available at the fingertips of any patient, whether at home, on the street, or at work.

Over the past decades, mHealth has developed rapidly. "This is reflected in the thousands of research articles listed on PubMed when searching for mHealth. In 2019, 4,880 results were retrieved, and in 2020, the search string resulted in 9,144 articles" (Rivera-Romero et al., 2023, p. 10). While the opportunities brought by mHealth are numerous, the challenges are equally significant. Issues such as data security, privacy, and the commodification of intimacy have become critical and urgent topics in academic, scientific, legal, and political circles (Aljedaani & Babar, 2021). Technical issues, whether medical or related to public health, still require better characterization, analysis, reflection, and new solutions (Istepanian & Al-Anzi, 2018).

In a recent literature review, Rivera-Romero et al. (2023) identified that the vast majority of mHealth services do not explicitly report the theoretical foundations that guide and support the conception, development, and implementation of their services. Therefore, while it is possible to identify theories and approaches being constructed, they are still in development—a process that, incidentally, requires time, accumulation, and, above all, interdisciplinary cooperation.

The effort of this chapter is to contribute in this direction. Drawing from the public health thinking inherent in Collective Health—*Saúde Coletiva* in Portuguese (SC)[1], we aim to synthesize the first version of a theoretical-conceptual framework

[1] Editors' notes: In Brazil, Collective Health is a field of study and practice that encompasses a holistic approach to health, integrating elements of public health, social medicine, and public health policy. It emphasizes the social determinants of health, addressing the broader socio-economic, cultural, and environmental factors that

(*Contd.*)

for mHealth. The goal is to contribute to the conception and development of this service in a way that incorporates the guiding principles of the Brazilian Unified Health System—*Sistema Único de Saúde* in Portuguese (SUS), particularly the principle of comprehensiveness (Pinheiro & Mattos, 2006). This effort is committed to strengthening and expanding the public nature of care and health attention practices in times of their digital transformation.

In general, mHealth has been considered under two aspects. First, as a customized health assistance service (Istepanian, 2022), capable of offering health actions to users according to their respective profiles. Second, as an important tool for mass data collection and extraction (Lupton, 2013), where data is gathered and stored globally at "explosive levels" (Istepanian & Al-Anzi, 2018). The usage value of this data can be utilized for information analysis and knowledge production in "public health" (Gall & Suzuki, 2019).

From the perspective of Collective Health, these two aspects can be understood through a holistic tool known as the "healthcare model"—*Modelos de Atenção* in Portuguese (MA) (Teixeira et al., 2014). This model refers to the different approaches adopted to organize and provide healthcare to individuals and populations within a respective health system (Souza & Bahia, 2014). It involves strategies that aim to guide and direct how health actions occur in the relationship between users-patients and professionals-service providers. MAs can vary according to the type of health service, the level of care (primary, secondary, tertiary), the specialty, and the population served. Some care models emphasize preventive medicine and health promotion, while others focus more on disease treatment and the provision of curative care.

One of the central issues concerning MAs revolves around comprehensiveness. This implies organizing/governing health services and actions in a way that integrates health promotion, disease prevention, treatment of illnesses and sufferings, and health recovery. It also means coordinating the hierarchical levels of care (primary, secondary, and tertiary). In these terms, comprehensiveness is made possible through a diversity of integration processes, involving various different tasks among the distinct health actions (Teixeira et al., 2014).

Among these, the integration between the fundamental poles of "health technologies" stands out: clinical technology, focused on a set of individual care actions, and sanitary technology, aimed at collective actions with a population scope (Mendes-Gonçalves, 1994; Schraiber et al., 1996). The more integrated the individual- and population-focused actions are, the more comprehensive the care practices will be. Therefore, an MA will be closer to achieving comprehensiveness. This integration remains a significant challenge for any MA (Teixeira et al., 2014).

---

impact health outcomes. Collective Health is closely aligned with the principles of the Brazilian Unified Health System (SUS), which include universal access, equity, and comprehensiveness of care. This approach advocates for the collective responsibility in promoting health and well-being, prioritizing community participation, and intersectoral collaboration to achieve health for all.

With this in mind, in this chapter, we argue that mHealth can be seen as a new condition of possibility for integrating the two poles of health technologies. This occurs, in our view, due to the automation of the feedback cycle between individual and population-level actions facilitated by Artificial Intelligence (AI) techniques embedded in mHealth applications (apps). In other words, as mHealth provides a digital service of individual assistance to the citizen, effectively *acting as* clinical technology, this same service co-produces with the user a set of data at the population level that contributes to sanitary technology. We refer to this as cybernetic integration in health. This presents a new condition of possibility for integrality, as it allows the value produced in addressing a health problem at the population level to participate in a customized manner in clinical-individual actions. These individual actions, in turn, can dynamically adapt to the needs and public health actions.

The chapter is divided into five parts, in addition to this introduction. In the first part, we revisit the Health Work Process Theory (HWPT) (Paim et al., 2017) to present our understanding of health technologies and characterize the two main poles: clinical technology and public health policy technology. Next, we briefly develop the target image of integrality in health, emphasizing the integration between these technologies. The third part, based on our theoretical approach to apps, the infrastructural or multisituated approach, characterizes apps as intermediaries in a digital ecosystem centered on cloud computing services. At the core of the chapter is the fourth part. Here, we first present the distinction between small data and big data, corresponding to clinical technology and sanitary technology. Then, we describe the automation of the feedback loop through the customization process executed by Health Recommendation Systems (HRS). In conclusion, we summarize our understanding.

Furthermore, the theoretical-conceptual considerations made here are relatively abstract and are not intended to be exhaustive. The aim is merely to present a framework that can contribute to the organization of future and shared investigations, or as Ricardo B. Mendes Gonçalves once said, serve as a "command of intelligibility for the apprehension of concrete practices" (Mendes-Gonçalves, 1994, p. 95). Therefore, if what is presented here is received by the reader as a plausible invitation for new contributions towards its improvement, we will have achieved our goal.

Health Technologies: The Two Structuring Poles

Promoting health, preventing risks and diseases, and caring for people who fall ill requires a heterogeneous set of health services. The term "health services" gains different meanings depending on the context in which it is used. Sometimes it appears as a synonym for "health delivery"—such as a consultation, a home visit, a vaccination, etc.—and sometimes it is used as a synonym for "health provider"—referring to a health post, a health center, an emergency unit, or a hospital (Paim, 2019, p. 190).

By opting for the meaning of "health delivery", we can reflect on and analyze the set of services in their various and heterogeneous interactions and possible combinations through the concept of the MA. One theoretical-conceptual way to define it is offered by the Theory of the Health Work Process (TPTS): "A model of care can be understood as a kind of logic or rationality that guides a given technological combination in health practices (...) activated to solve problems and meet health needs, both individual and collective" (Paim, 2019, p. 463).

This understanding sheds light on and analyzes health actions from the perspective of production relations and the work process (Mendes-Gonçalves, 1994; Schraiber et al., 1996). Thus, when we speak of health services as work processes, we consider the dialectic between the knowledge, techniques, and instruments that contribute to making these services happen, to produce and intervene in their work objects (Mendes-Gonçalves, 1994).

According to the TPTS, the production of health services always involves three procedural moments: the work as an activity, the knowledge and instruments used, and the objects taken as raw material for health work. These moments are all articulated in relation to the project underway, with a specific teleology. Hence, according to this approach, health technology does not merely refer to equipment or technical artifacts but refers, above all, to "the technical connections established within the work process between the operating activity and the work objects through those instruments. It refers to the configurations mutually assumed or assumable by the set of those three moments of the work process, given the objectives and purposes of the same" (Mendes-Gonçalves, 1994, p. 18).

Historically, since the demographic explosion caused by the Industrial Revolution within the context of modern states (Foucault, 1999; Rosen, 1993), it became necessary to address two poles of concern as health problems: "population bodies" and "individual bodies" (Foucault, 1999) or, as Geoffrey Rose (2001) stated, "sick populations" and "sick individuals". Both became targets of different actions that "established them as objects of knowledge and power" (Moraes, 2002).

The first sought the "apprehension of anatomical-physiological bodies collectively, and the establishment of such links between this collective and the environment" (Mendes-Gonçalves, 1994, p. 65). In this context, technologies based on collective explanations and interventions were structured and gained "social prominence, forming the sanitary aspect, the foundation of 19th century epidemiology" (Nemes, 2000, p. 53). The knowledge and instruments dedicated to population phenomena, that is, "which take the 'collective' health and disease process as their object of work" (Merhy, 2014, p. 21), thus shaped sanitary technology.

In this sense, paraphrasing Emerson Merhy (2004), a machine like a smartphone and its apps are not technologies in themselves but technological equipment/artifacts. They are expressions of a technology, which, in reality, consists of how certain knowledge seeks and uses the equipment as an instrument to achieve a specific objective. Speaking of technology according to this theory, therefore, means using "the unifying theoretical element", the common signifier "of the connections (the arrangement) established between the elements (the work object, the work instruments, the work activity) of a certain practice (the medical experience), at its internal level, and at the same time in relation to the social articulations that determine it" (Mendes-Gonçalves, 1994, p. 85).

Public Health Policy technology has been structured throughout history, especially around knowledge such as epidemiology and Public Health institutions. It understands health and disease as a population process, resulting from external aggressions that the individual biological body suffers from the environment. Its goal is to discover and isolate contagious sick individuals, sanitize the environment, destroy biological vectors, and protect the healthy (Massuda, 2008, p. 177). To achieve this, it developed intervention instruments on populations, such as epidemiological surveillance, sanitary surveillance, and special campaigns and programs. It also focuses on health education actions, control of infectious-contagious diseases, sanitation, and more recently, monitoring chronic-degenerative diseases, risk factor prevention, and health promotion (Campos, 2003).

The second pole focuses on the individualization of the normal and pathological at the level of the individual-biological human body, contained within the physical, chemical, and biological limits of the human body (Mendes-Gonçalves, 1994, p. 65). Also called medical technology, its object of intervention is the health and disease process from the perspective of the individual and the biological body, within a universe regulated by the concepts of normal and pathological (Merhy, 2014, p. 21). As Jairnilson Paim put it: "This body, bearer of needs translated as diseases, would be the object of intervention through work means, such as medical knowledge and its diagnostic and therapeutic instruments" (Paim, 2012, p. 463).

To achieve its objectives, such technology was primarily structured around official medicine, or biomedicine (Camargo Jr., 2005), enveloped in the clinical knowledge that operates medical assistance work. Its actions are centered on individual and specialized medical care, both outpatient and hospital-based, focusing on the treatment and rehabilitation of singular clinical cases (Campos, 2003). This means actions focused on individualized consultations, "dependent only on each doctor in front of their case" (Schraiber et al., 2023, p. 246), based on "an assistance project for each case, with clinical or surgical interventions, which the doctor advises on because they are being consulted, and the patient can choose to follow or not" (Schraiber et al., 2023, p. 246).

Although health technologies have undergone numerous reconfigurations and internal restructurings since the 18th century, the poles they focus on persist and remain central and unavoidable subjects of health services and actions in any MA. Consequently, discussing MA always involves reflecting on, proposing, and constructing combinations between sanitary-population technologies, which have a collective nature and are geared towards health surveillance, health education, sanitation, vector control, immunization, etc., and clinical-individual technology, centered on individualized diagnosis and therapy (both elective and emergency). In effect, we can say that the rationalities guiding a given combination or integration of these technologies are at the core of comprehensiveness in health.

To achieve its objectives, this technology was primarily structured around official medicine, or biomedicine (Camargo Jr., 2005), which is embedded in the clinical knowledge that operates medical assistance work. Its actions focus on individual and specialized medical care, both outpatient and hospital-based, aiming at the treatment and rehabilitation of singular clinical cases (Campos, 2003). These actions involve individualized consultations, "dependent only on each doctor in front of their case" (Schraiber et al., 2023, p. 246), based on "an assistance project for each case, with clinical or surgical interventions, which the doctor advises on because they are being consulted, and the patient can choose to follow or not" (Schraiber et al., 2023, p. 246).

Despite numerous reconfigurations and internal restructurings since the 18th century, the focal points of health technologies persist. They remain central and unavoidable subjects of health services and actions in any MA. Consequently, discussing MA always involves reflecting on, proposing, and constructing combinations between sanitary-population technologies, which are collective and geared towards health surveillance, health education, sanitation, vector control, immunization, etc., and clinical-individual technology, which focuses on individualized diagnosis and therapy (both elective and emergency). The rationalities guiding a given combination or integration of these technologies are at the core of comprehensiveness in health.

Comprehensiveness in Health: A Goal with Many Challenges

The clinical-and-public-health-policy proposal that underpins Brazil's SUS includes several guiding principles, among which are universality, equity, and comprehensiveness. The principle of universality focuses on efforts to ensure access to health services for all. Equity requires us to consider—politically, economically, culturally, technically, and ethically—what each person needs and what is possible according to their abilities and limitations. One of the central debates in the tradition of Collective Health about Models MAs has been precisely to reflect, formulate, and propose different forms of integration between health promotion actions, risk factor prevention, care for harm, and rehabilitation. In other words, it has dealt with comprehensiveness.

In Law No. 8.080 (1990), which underpins the SUS of Brazil, we see: "assistance to individuals through actions of promotion, protection, and recovery of health, with the integrated execution of care actions and preventive activities". This same law adopts the "comprehensiveness of care, understood as an articulated and continuous set of preventive and curative actions and services, both individual and collective, required for each case at all levels of complexity of the system". However, the specialized literature lacks an exact and consensual definition of what comprehensiveness would be (de Mattos, 2009, p. 16), as it "challenges us to know and do the 'what' and 'how' can be accomplished in health to universally meet everyone's needs" (Ayres, 2009, p. 14). For this reason, it may be the principle "most difficult to define and, yet, the one on which the greatest challenges rest for the effective consolidation of SUS and its lofty missions" (Mattos, 2009, p. 11).

A reasonable way to conceptualize comprehensiveness in health is provided by Mattos (2009). He suggests considering comprehensiveness in health as an 'image-objective'. Drawing on a contribution from the Chilean Carlos Matus[2], the concept of 'image-objective' aims to designate a certain configuration of a system or situation that we consider 'desirable'—managerially, socially, culturally, politically, and ethically. It serves as a normative horizon that "differs from a utopia because the actors who support it believe that such a configuration can be made real within a defined time horizon" (Mattos, 2009, p. 45).

The operational function of the 'image-objective' is to establish distinctions between what is aspired to be built and what already exists. The idea is to reject the reduction of reality to what already exists, to the here and now, in order to reveal what is inscribed in it as potential, in Aristotle's words, what has not yet actualized into a fact but can or cannot be realized, depending on causes and contingencies. Its epistemological, ethical, and political quality lies in the demand to consider, in the process of analysis, that knowledge necessarily exceeds the limits of our immediate thoughts about the experience itself. This forces us to ensure that both our consciousness of the existing and the experience of the objects themselves remain in continuous metamorphosis, committed to the becoming of reality. For this reason, the image-objective is never specific, and not ever tied to an empirically determined reality.

Given this, what is the image-objective of comprehensiveness in health? The starting point is to clearly distinguish between inclusiveness, as this guiding image, and integration, which refers to the various processes, the different tasks necessary for the health model to achieve comprehensiveness (Teixeira et al., 2014). In other words, extensiveness entails considering the diverse means and forms of integration that contribute to the continuous provision of services aiming to ensure the promotion, protection, cure, and rehabilitation of individuals and collectives.

Given this, following Mattos (2009) and Paim (2012), we can identify three facets that constitute the image-objective of comprehensiveness. The first facet

[2] Editors' notes: Carlos Matus (1931–1998) was a prominent Chilean economist and political scientist known for his work on public administration and strategic planning. He served as a minister under President Salvador Allende and later developed the Theory of Situational Planning, which emphasizes the importance of context and adaptability in the planning process. His contributions significantly influenced public policy and management practices in Latin America. Matus's work remains relevant for those studying governance, planning, and organizational theory.

refers to "integral medicine". As a critique of the fragmentation of medical specialties and the consequent Cartesian and reductionist decomposition of the body and human experience into different anatomic-physiological systems, integral medicine generally proposes a holistic or comprehensive approach to the patient and their environment. This approach focuses not only on the body as an integrated whole but also on all aspects of life, including emotional, social, and cultural dimensions. Unlike official medicine, which concentrates on treating specific symptoms or diseases, integral medicine seeks to address the underlying causes of health problems and promote overall balance in a person's life.

Thus, for it, comprehensiveness relates to good medical practice, "a form of professional conduct encompassing biological, psychological, and social dimensions" (Paim, 2012, p. 466). It is characterized by the refusal to reduce the patient to a set of "biological systems or apparatuses with lesions or dysfunctions" (Mattos, 2009, p. 52).

The second and third aspects refer to the organization and governance of health actions. The second aspect concerns the possibilities for integrating actions related to health promotion, disease prevention, treatment of illnesses and suffering, and health recovery/social reintegration, known as horizontal integration. The third aspect refers to the vertical dimension of integration, which aims to coordinate the hierarchical levels of care (primary, secondary, and tertiary) based on the nature of the problem at hand, ensuring continuity and longitudinal care. Additionally, these two aspects, as noted by Maia et al. (2010, p. 683), understand comprehensiveness "in a macro sense, by articulating health services into a complex network with other health services and institutions, not necessarily belonging to the health sector". These are the aspects of interest here, specifically the integration of the two technological poles mentioned earlier.

This polarization, it is important to mention, resonates with the "traditional dichotomy" between 'surveillance' and 'assistance' (Paim, 2003) observed in the history of health actions and services in Brazil and the efforts around the MAs proposed in the Brazilian context. Before the establishment of the SUS, until the late 1970s, health organizations in Brazil divided their actions and services between "sanitary assistance" provided by "public health" establishments, on one side, and "medical assistance" offered by Provident Medicine, on the other (Schraiber et al., 2023).

This division was not only present in administrative structures concerning budgetary, financial, and legal issues that limited or prevented an institutionally integrated functioning of the system. It also occurred in the health work process with its technologies. On one side, public health and its sanitary technology focused on the population with a more preventive and collective nature. On the other side, clinical technology focused on individual healthcare with curative individual actions, organized primarily in social security hospitals. This dichotomy resulted in a deep practical dissociation between them, with little or no articulation in service delivery. For example, the work of doctors and nurses did not connect with the work of the epidemiological surveillance team (Mota & Schraiber, 2011).

Furthermore, public health actions were not integrated with the needs, interests, and desires found in individual care (Campos, 2000).

This configuration began to change only in the late 1970s with the Brazilian Health Reform movement. Within this movement, the consolidation of the Collective Health field, the Primary Health Care agenda, and the establishment of the SUS in the 1980s were crucial. As Massuda (2008, p. 175) summarizes:

> "At the center of the propositions of the Health Reform movement and the Collective Health field are issues such as considering health as a citizenship right, reformulating health practices by expanding the understanding of the health-disease process beyond the biological component to include subjective and social issues, and re-establishing the integration between individual and collective health actions."

To a large extent, we can say that the term 'unified' in Brazil's Unified Health System (SUS) arose precisely due to the struggle for integration and unification of these poles. However, even though SUS was crucial in overcoming the prevailing institutional dichotomy, comprehensiveness has remained a challenge filled with tensions and divergences, whether technical, scientific, ethical, administrative, or political (Paim, 2003; Mattos, 2009).

Even today, the integration of the two technological poles continues to be one of the most significant challenges for any MA, featuring in various SUS legislations and guidelines, whether in the National Primary Care Policy (Brazil, Ministry of Health, 2018a) or the National Health Surveillance Policy (Brazil, Ministry of Health, 2018b). Ultimately, all emphasize that the absence or insufficiency of this integration distances the image-objective of comprehensiveness. Consequently, it hinders the realization of an MA that provides health services ensuring health promotion actions for defined population groups, health surveillance actions aimed at controlling risks and damages, care and recovery actions for sick individuals, early disease detection actions, diagnostic, treatment, and rehabilitation actions (Teixeira, 2011).

With all this in mind, we understand that mHealth can be seen as a new possibility for developing solutions directly aimed at resolving this historical and structural issue.

Apps as Intermediaries in the Digital Ecosystem

Smartphones have become a constant tool in our daily lives. Over the past decade, we've witnessed a rapid increase in smartphone users worldwide, reaching 7.2 billion users, with over 70% in low- and middle-income countries (Miller & Matviyenk, 2014). This technology has advanced, becoming more miniaturized and affordable, so that today's mobile phones, equipped with cameras, fingerprint sensors, motion sensors, internet connections, and more, can generate, store, and process individual and environmental data like never before in human history. This granular, real-time data, available 24 hours a day, is now used for various health actions.

Embedded in smartphones are other technical artifacts: applications (apps). As Miller & Matviyenk (2014) noted, since their market introduction in the late 2010s, apps have progressively become significant cultural, political, and economic forms for nation-states and the global political economy. Today, smartphones and thousands of mobile health apps are used worldwide by patients, doctors, and healthcare providers.

However, as Gerlitz, Helmond, Nieborg, and van der Vlist (2019) note, despite apps becoming increasingly embedded in the daily lives of billions worldwide—whether in our pockets, on our wrists, in our offices, factories, stores, homes, or cars—the study of apps is still in its early stages. A promising approach in this area follows the so-called "infrastructural turn" in digital media studies.

According to some proponents, this approach allows for greater sensitivity to issues of relational power, such as forms of economic-data-platform, and infrastructure power. Apps are not only situated in a technical or material sense but also in an economic and institutional sense. They can be related to other objects, devices, systems, infrastructures, clouds, and environments by both human and non-human actors (Gerlitz et al., 2019). Certainly, taking the infrastructural question seriously involves considering not only the relationships between apps but also their connections with other entities, agents, practices, systems, and structures—both digital and informational, as well as political, economic, social, and cultural. From a sociological perspective, this means understanding apps as sociocultural, sociotechnical, and technopolitical artifacts (Lupton, 2017).

In the immediate sense, apps have a relatively narrow definition, appearing as compact software designed to operate quickly and easily on devices like smartphones after being downloaded from a central distributor, such as an app store. However, these software applications are not merely digital commodities we purchase on the App Store or Google Play, similar to how we might buy a book on Amazon or *Mercakdo Livre*.

As Dieter, Gerlitz, Helmond, Tkacz, van Der Vlist, & Weltevrede (2019, p. 7) state:

> "downloading and installing apps through official app stores is regularly presented as a seamless experience, where users do not see the app being downloaded onto their devices; instead, they only experience the automatically installed version. In this way, app stores obscure the status of apps as concrete software objects" (cf. Morris & Elkins, 2015). To fight against this obfuscation, it is necessary to resituate apps outside their normative consumption context.

In other words, apps should be viewed as technical artifacts that prominently participate in the construction of an infinite network of simultaneous services.

These processes occur without users knowing exactly when, how, and where numerous connections and data flows happen. Apps are not autonomous, self-contained entities; they are distributed and decentralized cells, or as Dieter et al. (2019) describe, multisituated. They are multifunctional, systematically interacting with various stakeholders including components, objects, data, designers, users, owners, developers, governments, regulatory agencies, app stores, hardware and software infrastructures, public and private institutions, large corporations, and multilateral organizations. Each of these entities integrates and values apps in multiple simultaneous situations (Dieter et al., 2019).

Moreover, apps are multilocalized because they are intrinsically coupled to an emerging digital ecosystem operating on different scales—local, regional, national, and global (Dieter et al., 2019). Hence, apps are technical artifacts with different use values for various actors simultaneously, given their presence in diverse operational situations and locations, always interconnected.

As platforms have been defined in the literature as "the provision of computational infrastructures that allow multiple parties, such as users, developers, and advertisers, to develop the platform's data and functionalities" (Weltevrede & Janse, 2019). The approach used here builds on this notion of infrastructure or ecosystem and expands it to include apps (Weltevrede & Janse, 2019). This perspective offers decisive contributions to go beyond viewing the app as a simple passive object for users or merely a tool for a specific purpose. It emphasizes understanding the app as a privileged intermediary of a digital ecosystem that "combines and recombines data from heterogeneous and contingent sources simultaneously", making them available to a vast array of interested actors with different purposes, which in turn feed customized services back to the app (Weltevrede & Janse, 2019).

The planetary-scale computation that Benjamin Bratton (2016) speculated about a decade ago is now taking empirical shape within the emerging digital ecosystem that van der Vlist, Helmond, and Ferrari (2024) have termed "Big AI". This term first references the idea that, so far, "there is no AI without Big Tech". In other words, the transition of AI from a field of technical and scientific research and development to its phase of production, application, and commercialization integrated into the global economy and the daily lives of citizens, businesses, and governments—what the authors call the "industrialization of AI"—has been carried out by Big Tech companies.

Secondly, the term refers to the understanding that this industrialization is based on the deep interdependence between AI and "cloud computing", making the latter the backbone of the digital ecosystem (van der Vlist et al., 2024, p. 2). This means that AI and machine learning are not isolated products or services. They are integral components of a broader technology stack and an ecosystem of cloud infrastructure tools, products, and services. These various cloud offerings form the foundation for developing and deploying AI applications and solutions in companies across different sectors, each with its unique requirements and objectives (Idem, p. 7).

The digital ecosystem consists not only of a myriad of gadgets and sensors (Istepanian & Al-Anzi, 2018) that communicate with each other, establishing interobjectivity as Yuk Hui (2016) describes. More importantly, it includes data warehouses, data lakes, and data centers that handle volumes and varieties of data on a big data scale (Tecco and Zweig, 2017, July 10) and offer cloud computing services.

However, we should simplify this complexity in this chapter. For our purposes, it is reasonable to consider the digital ecosystem as a complex and opaque system (Gerlitz et al., 2019; Dieter et al., 2019), characterized primarily by the extraction, collection, storage, and manipulation of vast data sets to produce use values through cloud computing services, such as AI. Given this, we can view apps as intermediaries in this ecosystem, through which embedded AI executes its cloud-based services (Gerlitz et al., 2019; Dieter et al., 2019; van der Vlist et al., 2024).

Cybernetic Integration between Health Technologies

To present the cybernetic integration between public health policy and clinical technology, we will use an analytical framework based on the interaction between the user and the mHealth app that provides the service. Everything happens simultaneously for the user, but analytically, it is possible to break down the process into two axes for better presentation.

First, on the y-axis (Figure 1), we distinguish the scope and scale of data into small data and big data. When we look at the intimate and direct interaction between the user and the app's front-end, at the individual scope and scale, we refer to small data and small data'. When we consider the user's indirect interaction with the digital ecosystem via the back-end, at the population scope and scale, we refer to big data and big data'.

Second, on the x-axis, to characterize the automation of the feedback cycle, we use the metaphors of downstream and upstream flow.

The y-axis (Figure 1) emphasizes the scope and scale of the data, and the x-axis focuses on the automation flow, but both axes form the cybernetic integration between the two poles. Starting from the user, the process begins with data co-production during the interaction between the user and the app's front-end. Here, the small data acts as input, flowing downstream to construct the big data. Subsequently, after various actions performed by the AIs embedded in the app as cloud computing services, the big data undergoes a metamorphosis, forming a new use value at the population level, referred to as "big data". This, in turn, returns upstream as output in the form of customized digital assistance, creating a new individual-level use value, referred to as "small data".

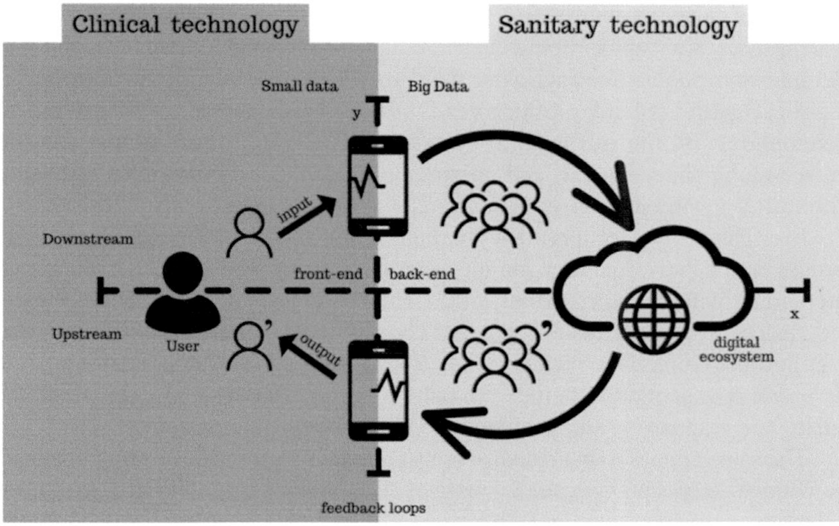

Figure 1: mHealth as cyber integration.

Taking on the Role of Health Technologies: The Small Data and the Big Data

According to the "relational view of data" proposed by Anne Beaulieu and Sabina Leonelli (2021), data do not speak for themselves. They do not represent reality or provide access to it without mediation. Instead, for data to represent something, "a process of investigation pursued by human beings in a specific context" is required (Beaulieu & Leonelli, 2021, p. 53). Consequently, "Objects only become data once you decide that they should be used as evidence for a particular claim" (Idem, p. 53). For example, a poem can simply be a piece of literature; but when it is used as evidence to claim that Marlowe is actually the author of Shakespeare's sonnets, that poem becomes data. Similarly, a photograph of a tree can be a beautiful reminder of a sunny holiday, but if it is taken as evidence of the tree's health on that specific day, that photograph becomes data. These everyday examples are no different from how data functions in Big Data contexts or in audit cultures where indicators dominate. A 'like' on Facebook only becomes data when it is recorded, collected, connected to the liked object and/or the liker's profile, and compared with other likes to serve as evidence for making a claim about popularity, social ties, or preferences (Idem, p. 55).

In our case, the claims concern the health-disease process, and for each claim, we will find a set of scientific, philosophical, political, ideological, and cosmological theories and approaches that support and justify them—or, as the TPTS states, we will find some technological knowledge that operationalizes them (Mendes-Gonçalves, 1994; Schraiber et al., 1996).

Historically, as mentioned earlier, sanitary technologies have utilized epidemiology and "epidemiological surveillance" to collect, manipulate, analyze, and interpret population data to support and/or execute actions with utility in "Public Health" (Mendes-Gonçalves, 1994; Schraiber et al., 1996). Clinical technologies, on the other hand, have traditionally followed biomedicine to collect, manipulate, analyze, and interpret individual data and provide assistance with utility for the patient.

In mHealth, we observe a cybernetic counterpart occurring. Small data focuses on the individual user, aiming to produce utility for the individual through interaction with the app's front-end. Centered around various technological knowledge—especially in the context of User Experience (UX) and User Interface (UI) design (Gomez-Hernandez et al., 2023)—digital assistance actions target the user's engagement with the app, tailored to each specific case, to which the patient can respond or not, analogous to clinical consultations.

The same applies to big data. It is constituted through various instrumentalized knowledge—especially in the context of data science and data analytics—for collecting, extracting, manipulating, and analyzing data in an aggregated manner (descriptive, diagnostic, predictive, or prescriptive), aiming to produce utility on a population scale, similar to "epidemiological surveillance". Therefore, following the Collective Health approach, we consider that in the context of mHealth, big data acts as public health policy technology, just as small data serves as clinical technology. Let's take a closer look, briefly.

Currently, mHealth services address a wide range of health problems. They span from endocrine, nutritional, or metabolic diseases to circulatory system diseases, infectious or parasitic diseases, neoplasms, pregnancy, childbirth, puerperium, nervous system diseases, musculoskeletal system diseases, respiratory system diseases, and even mental, behavioral, or neurodevelopmental disorders (Rivera-Romero et al., 2023).

mHealth is one of the most significant services in "digital health", primarily because it offers a privileged service for individualized and customized interaction (Istepanian, 2022). As Weltevrede and Jansen (2019) state, apps are central mechanisms in the individualization of digital services due to their integration into our daily routines. They are intimately linked to our habits, functioning as "intimate data environments" capable of providing customized experiences. They not only accompany our daily practices but also 'insinuate' themselves into our routines and habits.

The sociological literature on mHealth (Lupton, 2017) highlights how it supports and enhances initiatives for monitoring and quantifying bodies and 'selves', practices that have existed in healthcare for centuries. Self-tracking is becoming increasingly popular in health care (Lupton, 2013, 2017). Through self-tracking, individuals monitor various aspects of their health and well-being in real time. The clinical use values of these digital services are diverse, including helping identify signs or symptoms, providing early warnings related to chronic diseases, managing and adhering to medication and treatments, communicating changes in clinical status to health professionals, and many others—highlighting their similarities to clinical technology.

The primary goal of mHealth apps is to prompt users to change their perceptions of their bodies and habits, become aware of specific symptoms and diseases, learn about diagnostic tests, treatment options, and side effects, acquire self-management skills and informed decision-making, and modify their healthcare behaviors and practices (Cai et al., 2022). To achieve this, various UX and UI strategies are employed, including choice architectures, push notifications, nudges, affordances, chatbots, and gamification (Costanza-Chock, 2020). These diverse cybernetic techniques are used to engage users, encouraging them to interact, engage, and coproduce data.

In parallel, for these assistive services to be delivered to the user, mHealth promotes what is called datification (Van Dijck, 2014), focusing on the individually located health-disease process. Through datification, potentially all biological aspects and lifestyle choices of users are converted into data, for better or worse. These data are then manipulated, classified, modeled, and analyzed by the AI embedded in the apps. *Taking on the role* of "clinical gaze", the app collects and/or extracts various parameters, including anatomical and physiological metrics like body weight, body mass index, glucose levels, physical fitness, energy expenditure, body temperature, and heart rate. It also gathers information on sun exposure, dietary habits, alcohol consumption, sexual activity, smoking, sleep patterns, mental well-being, and menstrual cycles. Essentially, everything can be converted into bits.

Inspired by the proposal of Weltevrede and Jansen (2019), small data refers to the moment when data co-production occurs through direct interaction between the user and the front-end of the mHealth service. At this moment, individual-scope data (input) is co-produced, and the user receives back the clinical use value (output) they sought: small data—'small data'. For this to happen, for small data to return to the user as small data', the AI embedded in the apps must recombine and correlate it with numerous other data sources and information at the population scale and scope, in the big data.

Beneath the immediate interaction between user and front-end, in the back-end of the small data—'small data', lies the digital ecosystem we mentioned earlier. Cybernetic integration isn't possible without it. For instance, when a user utilizes mHealth services to self-monitor their fertility cycle or to track and manage their dermatological treatment, the app simultaneously functions as a "clinical gaze", collecting and/or extracting data and sending it downstream to the back-end. There, the digital ecosystem manipulates the data in an automated manner through the AI embedded in the app, aiming to produce new use values in big data.

There are many ways to define big data. Rob Kitchin and Gavin McArdle (2016) identified 26 different definitions in the scientific literature. Generally, big data is characterized by the abilities, developed over the past decades, to produce, store, and computationally analyze large volumes, varieties, and velocities of data. It's important to note that these capabilities are impossible for a human cognitive system to achieve. As Sabina Leonelli (2022) states, "who among us truly understands the difference between a trillion and a quadrillion?" (p. 22)—let alone can memorize and correlate them within minutes or seconds.

A significant body of literature has rightly highlighted the economic, political, social, and ethical issues of big data as a digital ecosystem of massive data extraction and collection (Couldry & Mejias, 2019; Zuboff, 2019). We must not lose sight of these concerns, nor can we ignore the numerous layers this ecosystem possesses—from content layers (texts, videos, images, audios, etc.) to physical layers (Wi-Fi, fiber optics, etc.). Additionally, we must acknowledge the new forms of work it demands, such as "AI data work" (Muldoon et al., 2024). As Leonelli (2022, p. 25) states,

> "Instead of attempting to define Big Data in terms of physical characteristics and quantity, I propose characterizing it based on how it is used. Big Data consists of data from different types and sources that are related, often in digital formats and ways that lend themselves to machine learning, thereby producing new analysis and knowledge procedures."

Not surprisingly, in the health context, some experts argue that "the next great frontier of epidemiology will be the analysis of large databases" (Chiavegatto Filho, 2015, p. 326). Similarly, we are witnessing the emergence of fields like 'infodemiology' (Eysenbach, 2011) and "digital epidemiology" (Salathé, 2018). Gall and Suzuki (2019), for instance, claim that the volume, variety, and velocity with which data is manipulated in big data represent forms of health surveillance that traditional surveillance does not possess. These forms allow for the production of risk calculations, forecasts, and predictions about users' behaviors, preferences, intentions, and tendencies that cannot be produced in the traditional way.

Thus, as the authors state, big data:

> "can be used at all three levels of health promotion and disease prevention —research, surveillance, and intervention—by: (a) allowing a more precise identification of at-risk populations, through a more comprehensive understanding of human health and disease, including the interaction between genetic, lifestyle, and environmental determinants of health; (b) enabling better surveillance of both communicable and non-communicable diseases; and; (c) facilitating better targeted strategies and interventions to improve health promotion and disease prevention" (Gall & Suzuki, 2019).

Consequently, one can produce new use values in the population scope, big data—'big data', *taking on the role* of sanitary technology. In fact, in the words of some pioneers in the field, Rob Kitchin and Tracey Lauriault: "individually, databases provide limited views of people, but they gain power when combined" (2014, p. 12). Or as Cedric Durand says, this is what big data is, "a 'sea of data' where algorithms extract a surplus that emanates from individual actions but, in the process of aggregation, transcends them and returns to them metamorphosed" (2020, p. 54).

Through interaction with an app designed for dermatological assistance, a user can produce photos of a suspicious spot on their skin. These photographs are then extracted and/or collected to function as data. Sent to the digital ecosystem, the AI embedded in the app manipulates these data to produce new use values, according to the scope of interest. An epidemiological perspective will focus on the incidence and/or prevalence of such spots in a given population. An individual perspective, a clinical one, will interpret the spot to suggest an anatomopathological diagnosis to the user.

However, the cybernetic integration promoted by mHealth becomes evident only when we consider that the beginning and end of the feedback cycle appear as small data—'small data'. In mHealth, everything starts and ends *taking on the role of* clinical technology. The small data flows downstream to the big data—'big data', returning upstream as customized assistance to the user, metamorphosed as 'small data'. To understand this, we need to look more closely at the x-axis of our framework.

Automation of the Feedback Cycle

Customization, as opposed to the "one-size-fits-all" approach, involves adapting mHealth services to the needs and preferences of each individual user. This adaptation occurs through feedback cycles that function as circular causality between the app (front-end and back-end) and the user—a cybernetic process where outputs return to the system as inputs, ideally continuously. To illustrate this, we will use one of the main AI techniques for service customization, which is now widely explored in mHealth: Health Recommender Systems (HRS) (Sun et al., 2023).

Recommender systems (RS) in general became popular at the beginning of the 21st century, especially in the context of online shopping, with the aim of recommending items of interest to users of a particular store. Today, they are present in social networks, search engines, apps, etc. Anyone who has used any streaming service, for example, has interacted with a recommender system. The RS functions as a stack of AI-embedded software in the app, specialized in delivering real-time individual recommendations based on automated user data analysis, predicting their needs, interests, and preferences. In essence, it delivers recommendations tailored to each personal device based on the user profile.

Traditionally, recommendations are given based on the user's previous searches or purchases. Based on the data produced by the users themselves—their preferences, contacts, interactions, views, virtually all the information they co-produce with the platform—the RS profiles the user and delivers "personalized recommendations". In the case of HRS, the scheme is similar, although modeled to recommend other items based on other data and information.

Currently, HRS can be found in services focused on asthma, cancer, vascular disease, depression, diabetes, lifestyle, Alzheimer's, obesity, sexual and reproductive health, sleep problems, smoking, among others (Cai et al., 2022). As Cai et al. (2022) state, the HRS is a specialized system that "provides users with personalized health information", capable of automatically identifying and recommending appropriate materials (such as diagnostic recommendations, health insurance, treatment methods based on clinical pathways, and alternative medicines) to users based on their specific health conditions and needs (p. 2). Consequently, it "presents an opportunity for the healthcare sector to transition from a traditional paradigm to a more personalized paradigm" (p. 2).

Based on the survey conducted by these authors from dozens of studies, we can say that HRS operates in five dimensions: "health domains", "users", "recommendation items", "recommendation technologies", and "evaluation systems" (Cai et al., 2022). The first dimension refers to health issues, and the last refers to methods used to evaluate the performance of HRS. In this chapter, we are interested in examining the other three dimensions, as they allow us to observe the automation of the feedback cycle and, consequently, the cybernetic integration between clinical technology and public health technology.

The "users" dimension concerns the target users who coproduce the small data and receive the recommendations. Without small data, the HRS cannot provide personalized recommendations. It also refers to "user models", which are profiles designed based on attributes related to a specific health issue—such as sex, age, body mass, location, occupation, marital status, height, family history, routine medication, physical activity frequency, diet, clinical data, etc. User modeling involves identifying which user characteristics are relevant to a service and designing a system structure to capture these attributes (Idem, p. 9). In other words, an app for diabetes patients builds a "user model" based on the determinants and risk factors of this disease, while an app for fertility self-monitoring builds its model based on the specificities of women's health, and so on, each with its own model.

"Recommendation items" refer first to numerous information sources that correlate with small data downstream to shape big data and produce new usage values. Examples include government databases (demographic, epidemiological, socioeconomic), specialized academic articles, books, guidelines, databases, and data from patients with similar health problems found on social media, patient websites, and electronic health records (Cai et al., 2022). Secondly, and most importantly, the items refer to recommendations directed upstream to the target user. This small data', which appears as output to the user, can include educational content related to the health issue, medication guidance, physical activity and diet recommendations, referrals to specific services, motivational messages, and suggestions for 'friendships' with patients in similar conditions, serving the role of clinical technology.

In the context of mHealth, customization has so far been aimed at five main purposes (Rivera-Romero et al., 2023): communication, targeting both healthcare professionals and patients; rehabilitation, supporting patients in their recovery practices; health promotion, specifically to change or maintain certain behaviors for healthy habits; and self-management, enabling patients to monitor and manage their own conditions—the clinical scope usage values.

At the core are the cybernetic operators of the interaction between "users" and "recommendation items", the "recommendation technologies". These technologies use various AI techniques, such as collaborative filtering, content-based filtering, knowledge-based filtering, comparison between techniques, and especially the most used, the hybrid recommender system, which combines the previous methods in different ways (Sun et al., 2023). Regardless of the technique, it operates what we call the automation of the feedback cycle between clinical and sanitary technology. It collects and/or extracts data, consolidates, and manipulates them to be transformed and then returned to users as a customized mHealth service.

The RS's ability lies in executing the cycle in an automated manner: small data—big data—'big data'—'small data'.

In this process, the data co-produced in the longitudinal experience of the individual user (small data) flows downstream as input to the digital ecosystem. This ecosystem aggregates and manipulates the data as big data. Subsequently, it returns upstream with added use values ('big data'), reaching the final user (output) as small data'. It is crucial to note that the data flows downstream only to return upstream to the same source, small data—'small data'. In this case, the new use value represented by 'small data' consists of customized "recommendation items" based on the user's profile in the context of big data—'big data' manipulation. Downstream, small data serves as raw material for big data—'big data', and upstream, it appears as the final product, as a service, 'small data'.

If the small data flows downstream to the digital ecosystem, it's because the use value of small data per se tends to zero. mHealth can only deliver use value to the user if the small data is transformed into 'small data', a customized recommendation item. This transformation occurs only after it has been manipulated from data within the population scope, big data—'big data'.

In other words, small data is collected and/or extracted to be subsequently aggregated, recombined, fed back, correlated, etc., with other data (big data—'big data'), and only then does it return as small data', as a customized mHealth service. There are various techniques to manipulate small data within big data—'big data' to profile the target user (Saxena & Saxena, 2020), finding patterns that allow reasonably precise calculations of future behaviors and events. Regardless of the specific technique, what interests us in this chapter is that the goal of SRS is the delivery of customized outputs.

At the end, the AI embedded in the apps as cloud computing services operate the small data—big data—'big data'—'small data' cycle, continuously and in real time, almost like an endless loop. Therefore, in mHealth, the clinical use value arises from the customization process carried out and delivered by SRS based on sanitary technology. This demonstrates cybernetic integration.

It's crucial to remember that the greater the quantity and quality of small data for manipulation at the big data—'big data' scale, the more customized the small data' will be. In other words, the more the user interacts with mHealth, the better the AI algorithms will be calibrated to deliver a customized service. This is how we see two types of customizations: static and dynamic. Static customization refers to a "situation where an algorithm personalizes using only static data, such as a nickname or nationality, which doesn't change during use". Dynamic customization "represents algorithms that use real-time data or data collected during use, like physical activity patterns or location" (Rivera-Romero et al., 2023, p. 3). Consequently, the more dynamic (Montaner et al., 2003) the feedback cycle is, the more integrated health technologies will be. The dynamic nature of the cycle can reflect the updated health status of the patient.

Conclusion

In this chapter, we argued that mHealth offers a new (cybernetic) condition for integrating the two foundational poles of health technologies: clinical technology and sanitary technology. Consequently, it directly impacts the objective image of comprehensiveness in health.

This understanding rests on the fact that mHealth converges both individual health actions and population health actions within the same device. After the conceptual foundation, the presented framework aimed to demonstrate that the production of clinical use value in the context of mHealth occurs within an automated feedback cycle. The processing axes (x and y) integrate small data and big data, showing that in mHealth, clinical and sanitary technologies integrate within the same technical artifact.

As a provisional conclusion of this ongoing analysis, mHealth directly contributes to longitudinal individual assistance while adapting to changes in the user's clinical condition and sanitary context. In other words, it offers individual assistance without disregarding the socio-sanitary situation, integrating efforts of adherence and engagement in individual clinical treatment and sanitary policy simultaneously.

All these considerations highlight the importance of integrating health technologies and emphasize the need for interdisciplinary approaches to develop and deepen theoretical and conceptual frameworks capable of addressing the complexity of challenges and opportunities in the field of "digital health". This is especially true with the ethical-political horizon of strengthening public health services that are free, secular, and universal.

Always keeping in mind that caring means participating in the socially shared construction of happiness projects that continuously create and emerge from our encounters in the world, each time producing new horizons for new possible encounters. Caring means ensuring that every experience presenting as an obstacle to what we want to make of our lives, our life in the polis and the polis in us, is understood and transformed in the way that seems most just, productive, and beautiful to us. Negatively, as critique and resistance to practices that distance us from our legitimate aspirations, or positively, in the proposition of alternatives, care and comprehensiveness are congenial ideas in their vast ambitions, generous to the limit of utopia, and therefore indispensable (Ayres, 2009, p. 12).

References

Aljedaani, B., & Babar, M.A. (2021). Challenges with developing secure mobile health applications: Systematic review. *JMIR mHealth and uHealth*, 9(6), e15654.

Ayres, J.R.D.C.M. (2009). Organização das ações de atenção à saúde: Modelos e práticas. *Saúde e Sociedade*, 18, 11–23.

Beaulieu, A., & Leonelli, S. (2021). *Data and Society: A Critical Introduction*. Los Angeles, CA: Sage.

Brasil. Ministério da Saúde. (2018a). Secretaria de Atenção à Saúde. Secretaria de Vigilância em Saúde. *Guia Política Nacional de Atenção Básica – módulo 1: integração atenção básica e vigilância em saúde*. Brasília: Ministério da Saúde. http://bvsms.saude.gov.br/bvs/publicacoes/guia_politica_nacional_atencao_basica_integracao_atencao_basica_vigilancia_saude_modulo_1.pdf.

Brasil. Ministério da Saúde. (2018b). Conselho Nacional de Saúde. *Resolução MS/CNS nº 588, de 12 de julho de 2018. Fica instituída a Política Nacional de Vigilância em Saúde (PNVS), aprovada por meio desta resolução*. Diário Oficial da República Federativa do Brasil, Brasília (DF), 2018, ago 13; *Seção*, 1, 87. https://www.gov.br/saude/pt-br/assuntos/saude-de-a-a-z/p/politica-nacional-de-vigilancia-em-saude.

Bratton, Benjamin H. (2016). *The Stack: On Software and Sovereignty*. MIT Press.

Cai, Y., Yu, F., Kumar, M., Gladney, R., & Mostafa, J. (2022). Health recommender systems development, usage, and evaluation from 2010 to 2022: A scoping review. *International Journal of Environmental Research and Public Health*, 19(22), 15115.

Camargo Jr, Keneth, R.D. (2005). A biomedicina. *Physis: Revista de Saúde Coletiva*, 15, 177–201.

Campos, Carlos E.A. (2003). O desafio da integralidade segundo as perspectivas da vigilância da saúde e da saúde da família. *Ciência & Saúde Coletiva*, 8, 569–584.

Campos, Gastão. W.D.S. (2000). Saúde pública e saúde coletiva: Campo e núcleo de saberes e práticas. *Ciência & Saúde Coletiva*, 5, 219–230.

Chiavegatto Filho, & Alexandre, D.P. (2015). Uso de big data em saúde no Brasil: Perspectivas para um futuro próximo. *Epidemiologia e Serviços de Saúde*, 24, 325–332.

Costanza-Chock, S. (2020). *Design Justice: Community-led Practices to Build the World's We Need*. Cambridge, MA: MIT Press.

Couldry, N., & Mejias, U. (2019). *The Costs of Connection: How Data is Colonizing Human Life and Appropriating It for Capitalism*. Stanford, CA: Stanford University Press.

Deloitte. (2014) Deloitte centre for health solutions. *Healthcare and Life Sciences Predictions 2020: A Bold Future*. Deloitte LLP, UK, 2014. https://www2.deloitte.com/content/dam/Deloitte/global/Documents/Life-Sciences-Health-are/gx-lshc-healthcare-and-lifesciences-predictions-2020.pdf.

Dieter, M., Gerlitz, C., Helmond, A., Tkacz, N., van Der Vlist, F.N., & Weltevrede, E. (2019). Multi-situated app studies: Methods and propositions. *Social Media & Society*, 5(2), 1–15.

Durand, Cédric. (2020). Tecnofeudalismo: la nueva gleba digital. *In*: *Viento sur: Por una izquierda alternativa*, 173, 51–61.

Eysenbach, G. (2011). Infodemiology and Infoveillance: Tracking online health information and cyberbehavior for public health. *Am. J. Prev. Med.*, 40/5 (Suppl 2), S154–S158.

Foucault, Michel. (1999). Aula de 17 de março de 1976. *In*: *Em Defesa da Sociedade. Curso no Collège de France* (*1975–1976*) (pp. 285–319). São Paulo: Martins Fontes.

Gall, C., & Suzuki, E. (2019). *5 Big Data: A New Dawn for Public Health? Health in the 21st Century*, 146. https://www.oecd-ilibrary.org/sites/f24cb567-en/index.html?itemId=/content/component/f24cb567-en.

Gerlitz, C., Helmond, A., Nieborg, D.B., & van der Vlist, F.N. (2019). Apps and infrastructures: A research agenda. *Computational Culture: A Journal of Software Studies*, 7. http://computationalculture.net/apps-and-infrastructures-a-research-agenda/.

Gomez-Hernandez, M., Ferre, X., Moral, C., & Villalba-Mora, E. (2023). Design guidelines of mobile apps for older adults: Systematic review and thematic analysis. *JMIR Mhealth Uhealth*, 11, e43186.

Hui, Yuk. (2016). *On the Existence of Digital Objects*. Minneapolis: University of Minnesota Press.

Istepanian, R.S., & Al-Anzi, T. (2018). mHealth 2.0: New perspectives on mobile health, machine learning, and big data analytics. *Methods*, 151, 34–40.

Istepanian, R.S. (2022). Mobile health (mHealth) in retrospect: The known unknowns. *Int. J. of Environmental Research and Public Health*, 19(7), 3747.

Kitchin, R., & Laurialt, T. (2014). Towards critical data studies: Charting and unpacking data assemblages and their work. *In*: Thatcher, J., Eckert, J., Shears, A. (Eds.), *Thinking Big Data in Geography*. New Regimes, New Research. Lincoln/London: University of Nebraska Press.

Kitchin, R., & McArdle, G. (2016). What makes Big Data, Big Data? Exploring the ontological characteristics of 26 datasets. *Big Data & Society*, 3(1).

Lei nº 8.080, de 19 de setembro de 1990. (1990), *Diário Oficial da União*. Available from: https://www.planalto.gov.br/ccivil_03/LEIS/L8080.htm.

Leonelli, Sabina. (2022). *A pesquisa científica na era do Big Data: Cinco maneiras que mostram como o big data prejudica a ciência, e como podemos salvá-la*. Rio de Janeiro: Fiocruz.

Lupton, Deborah. (2013). Quantifying the body: Monitoring and measuring health in the age of mHealth technologies. *Critical Public Health*, 23(4), 393–403.

Lupton, Deborah. (2017). *Digital Health: Critical and Cross-disciplinary Perspectives*. Book Series: Critical Approaches to Health (1st Edn.). London: Routledge.

Maia, Christiane, Guilhem, Dirce, & Lucchese, Geraldo. (2010). Integração entre vigilância sanitária e assistência à saúde da mulher: um estudo sobre a integralidade no SUS. *Cadernos de Saúde Pública*, 26, 682–692.

Massuda, Adriano. (2008). Práticas de saúde coletiva na atenção primária em saúde. *In*: Brasil, Ministério da Saúde. *Manual de práticas de atenção básica: Saúde ampliada e compartilhada* (pp. 179–205). São Paulo: Hucitec.

Mattos, Ruben A. (2006). Os sentidos da integralidade: Algumas reflexões acerca de valores que merecem ser defendidos. *In*: Pinheiro, Roseni & Mattos, Ruben (Eds.),

Os sentidos da integralidade na atenção e no cuidado à saúde (4ª edn.) (pp. 39–64). Rio de Janeiro: Uerj/IMS, Abrasco.
Mendes-Gonçalves, Ricardo Bruno. (1994). *Tecnologia e organização social das práticas de saúde. Características tecnológicas do processo de trabalho em saúde na rede estadual de centros de saúde de São Paulo*. São Paulo: Hucitec.
Merhy, Emerson, E. (2004). O ato de cuidar: A alma dos serviços de saúde. *In*: Brasil. Ministério da Saúde. *SUS Brasil: cadernos de textos*. Brasília: Ministério da Saúde,108–137.
Merhy, Emerson, E. (2014). *São Paulo de 1920 a 1940: A Saúde Pública como Política: Os movimentos sanitários, os modelos tecnoassistenciais e a formação das políticas governamentais*. Porto Alegre: Rede UNIDA.
Miller, P.D., & Matyivenko, S. (2014). *The Imaginary App*. Cambridge, MA: MIT Press.
Montaner, M., López, B., & De La Rosa, J.L. (2003). A taxonomy of recommender agents on the Internet. *Artif. Intell. Rev.*, 19, 285–330.
Moraes, Ilara Hammerli S. de. (2002). *Política, tecnologia e informação em saúde: A utopia da emancipação*. Salvador: Casa da Qualidade Editora.
Mota, André, & Schraiber, Lilia, B. (2011). Atenção primária no sistema de saúde: Debates paulistas numa perspectiva histórica. *Saúde e Sociedade,* 20, 837–852.
Muldoon, J., Cant, C., Wu, B., & Graham, M. (2024). A typology of artificial intelligence data work. *Big Data & Society*, 11(1).
Nemes, Maria I.B. (2000). Ação programática em saúde: Recuperação histórica de uma política de programação. *In*: Schraiber et al. (Eds.), *Programação em saúde hoje* (2 Ed.) (pp. 65–116). São Paulo: Editora Hucitec.
Paim, Jairnilson. S., & de Almeida Filho, Naomar A. (2023). *Saúde coletiva: Teoria e prática*. Rio de Janeiro: MedBook.
Paim, Jairnilson S. (2003). Epidemiologia e planejamento: A recomposição das práticas epidemiológicas na gestão do SUS. *Ciência & Saúde Coletiva*, 8, 557–567.
Paim, Jairnilson S. (2012). Modelos de Atenção à Saúde no Brasil. *In*: Giovanella L, Escorel, S., Lobato L.V.C, Noronha, J.C., Carvalho, A.I. (Eds.), *Políticas e Sistema de Saúde no Brasil* (2ª Ed.) (pp. 459–491). Rio de Janeiro: Fiocruz, Cebes.
Paim, Jairnilson S. (2017). Da teoria do processo de trabalho em saúde aos modelos de atenção. *In*: Mendes-Gonçalves, R.B. Saúde (Ed.), *Sociedade e História* (pp. 257–297). São Paulo: Hucitec.
Paim, Jairnilson S. (2019). *SUS: Sistema Único de Saúde: Tudo o que você precisa saber*. São Paulo: Atheneu.
Pinheiro, Roseni, & Mattos, Ruben Araújo de. (2006). *Os sentidos da integralidade na atenção e no cuidado à saúde* (6. Ed.). Rio de Janeiro: IMS/UERJ – CEPESC – ABRASCO.
Portela, Luis E., & Teixeira, Carmem F. (2012). Epidemiologia e gestão de serviços de saúde. *In*: Almeida Filho, N., & Barreto, M.L. (Eds.), *Epidemiologia & saúde: Fundamentos, métodos, aplicações* (pp. 2073–2106). Rio de Janeiro: Guanabara Koogan.
Rivera-Romero, O., Gabarron, E., Ropero, J., & Denecke, K. (2023). Designing personalized mHealth solutions: An overview. *Journal of Biomedical Informatics*, 146(2023), 104500.
Rose, Geoffrey. (2001). Sick individuals and sick populations. *International Journal of Epidemiology*, 30(3), 427–432.
Rosen, George. (1993). *A History of Public Health* (Expanded Edn.). Baltimore: Johns Hopkins University Press.

Salathé, Marcel. (2018). Digital epidemiology: What is it, and where is it going? *Life Sciences, Society, and Policy*, 14(1).

Saxena, Mohit, & Saxena, Ankur. (2020). Evolution of mHealth eco-system: A step towards personalized medicine. *In: International Conference on Innovative Computing and Communications: Proceedings of ICICC 2019,* 1, 351–370. Singapore: Springer.

Schraiber, L.B., Nemes, M.I.B., & Mendes-Gonçalves, R.B. (1996). *Saúde do Adulto: Ações e programas na unidade básica*. São Paulo: Hucitec.

Schraiber, Lilia B., Vilasbôas, Ana Luiza Q., & Nemes, Maria I.B. (2023). Programação em saúde e organização das práticas: Possibilidades de integração entre ações individuais e coletivas no Sistema Único de Saúde. *In*: Paim, J.S., & Almeida-Filho, N.A. (Eds.), *Saúde Coletiva: Teoria e prática*. Rio de Janeiro: MedBook.

Souza, L.E.P.F., & Bahia, L. (2014). Componentes de um sistema de serviços de saúde: População, infraestrutura, organização, prestação de serviços, financiamento e gestão. *In*: Paim, J.S., & Almeida-Filho, N.A. (Eds.), *Saúde Coletiva: Teoria e prática* (pp. 49–66). Rio de Janeiro: MedBook.

Sun, Y., Zhou, J., Ji, M., Pei, L., & Wang Z. (2023). Development and evaluation of health recommender systems: Systematic scoping review and evidence mapping. *J. Med. Internet Res.*, 25, e38184.

Tecco, H. & Zweig, M. (2017, July 10). *Digital Health Funding 2017 Midyear Review: A Record Breaking First Half.* https://rockhealth.com/insights/2017-midyear-funding-review-a-record-breaking-first-half/.

Teixeira, C.F., Vilasbôas, A.L., & Paim, J.S. (2014). Modelos de atenção à saúde no SUS: Transformação, mudança ou conservação. *In*: Paim, J.S., & Almeida-Filho, N.A. (Eds.), *Saúde Coletiva: Teoria e prática* (pp. 287–301). Rio de Janeiro: MedBook.

Teixeira, Carmem. (2011). Os princípios do sistema único de saúde. *In*: *Conferências Municipal e Estadual de Saúde*; 2011; Salvador-BA. Salvador. https://edisciplinas.usp.br/pluginfile.php/3023433/mod_resource/content/4/OS_PRINCIPIOS_DO_SUS.pdf.

Van der Vlist, F.N., Helmond, A., & Ferrari, F.L. (2024). Big AI: Cloud infrastructure and the industrialization of artificial intelligence. *Big Data and Soc.*, 11(1), 1–16.

Van Dijck, José. (2014). Datafication, dataism, and dataveillance: Big Data between scientific paradigm and ideology. *Surveillance and Society*, 12(2), 197–208.

Weltevrede, E., & Jansen, F. (2019). Infrastructures of intimate data: Mapping the inbound and outbound data flows of dating apps. *Computational Culture*, 7, 1–37. http://computationalculture.net/infrastructures-of-intimate-data-mapping-the-inbound-and-outbound-data-flows-of-dating-apps/.

World Health Organization (WHO). (2011). mHealth: New horizons for health through mobile technologies. *In*: *Global Observatory for eHealth Series* (Vol. 3). Geneva. https://iris.who.int/handle/10665/44607.

Zuboff, Shoshana. (2019). *The Age of Surveillance Capitalism: The Fight for a Human Future at the Newfrontier of Power*. New York: Public Affairs.

8 | Explainable Artificial Intelligence

Maíra Araújo de Santana*, Giselle Machado Magalhães Moreno and Wellington Pinheiro dos Santos

Artificial Intelligence (AI) is revolutionizing various sectors of society, particularly through its role in decision-making. In healthcare, AI has the potential to play a crucial role, from assisting in diagnosis to personalizing treatments. However, its lack of transparency and explainability presents ethical and practical challenges. Understanding and trusting automated decisions in healthcare is fundamental to ensuring patient safety and efficacy. Therefore, developing methods to make AI systems more explainable and transparent is essential. Explainable Artificial Intelligence (XAI) emerges as a response to this need, aiming to provide explanations for AI decisions and internal processes. This allows developers and, more importantly, users to understand how decisions are made based on input data and what factors influence them. Numerous studies have focused on developing techniques and approaches to make AI more explainable. This chapter presents a narrative review of recent work related to incorporating XAI techniques in healthcare. The studies are grouped according to their central application, encompassing XAI in diagnostic support, personalization of procedures, and health management.

Introduction

Artificial Intelligence (AI) is fundamentally transforming various areas of society, with one of its most significant applications being decision support. AI is praised for its ability to handle complex, varied, and large-scale data, providing pathways for more informed, precise, and efficient decision-making. However, it's important to note that using AI in decision-making also presents challenges, such as ensuring transparency, explainability, and ethics in increasingly automated decisions.

In healthcare, AI adoption is growing. It is used in Public Health, including digital epidemiology, environmental surveillance, geospatial health, and public

*Corresponding author: maira.araujosantana@ufpe.br

health policies, as well as assisting doctors and healthcare professionals in interpreting exams, diagnosing diseases, predicting clinical outcomes, and recommending personalized treatments. Machine learning-based AI systems can analyze medical images, such as CT scans and magnetic resonance imagings (MRIs), identifying subtle patterns that may be missed by the human eye. This not only speeds up the diagnostic process but can also improve the accuracy and quality of results.

The challenges accompanying AI use in healthcare emphasize the need for ethical clarity and transparency, which is crucial for the acceptance of new technologies and solutions by professionals, managers, and health agents. Considering that these actors make critical decisions that directly impact patients' lives, it is essential for them to understand how AI tools work. This understanding directly influences their trust in the recommendations and diagnoses generated by these systems.

The lack of explainability in AI can lead to several challenges and concerns. For example, in critical applications like medical diagnostics, it's essential for professionals and patients to understand how AI arrived at a specific conclusion or recommendation. Beyond understanding decisions, transparency in AI operations offers a greater possibility for: (i) detecting bias and error, as healthcare professionals can identify inappropriate patterns or inconsistencies in the results and suggested interventions, allowing for algorithm adjustments to ensure better decisions; (ii) enhancing clinical practice, as professionals use the relationships established by the system to construct more complex and potentially innovative reasoning about pathological processes, disease courses, and health practice in general. By enabling healthcare professionals to understand the functioning of these tools, it promotes a more ethical, safe, and effective use of AI in clinical practice, always focusing on the patients' well-being and health.

In the pursuit of greater transparency in AI systems, Explainable Artificial Intelligence (XAI) emerges as a fundamental concept in the AI field. XAI refers to the capability of these systems to provide understandable explanations about their decisions and internal processes. In other words, explainable AI aims to make the functioning of AI algorithms transparent, allowing users and their creators to understand how decisions are made and what factors influence these decisions.

The need for explainable AI arises from the complexity of AI models, especially in advanced techniques like deep neural networks and complex machine learning algorithms. These models can be highly effective in tasks such as pattern recognition, natural language processing, and decision-making, but they are often considered "black boxes" due to the lack of transparency in their internal workings.

Figure 1, introduced by Salih et al. (2023), displays some of the most common AI algorithms on a bidimensional plane of explainability versus performance. In it, the authors point out that deep networks often show higher performance but are associated with lower explainability, while decision tree-based models demonstrate greater explainability but lower performance. The chart also indicates the location of an ideal model, combining high performance and high explainability. It's important to note that this survey was conducted based on a literature review related to AI applied to cardiac imaging. Thus, although performances may vary depending on the problem observed, it is evident that current models alone do not yet fulfill the proposed function for the ideal model.

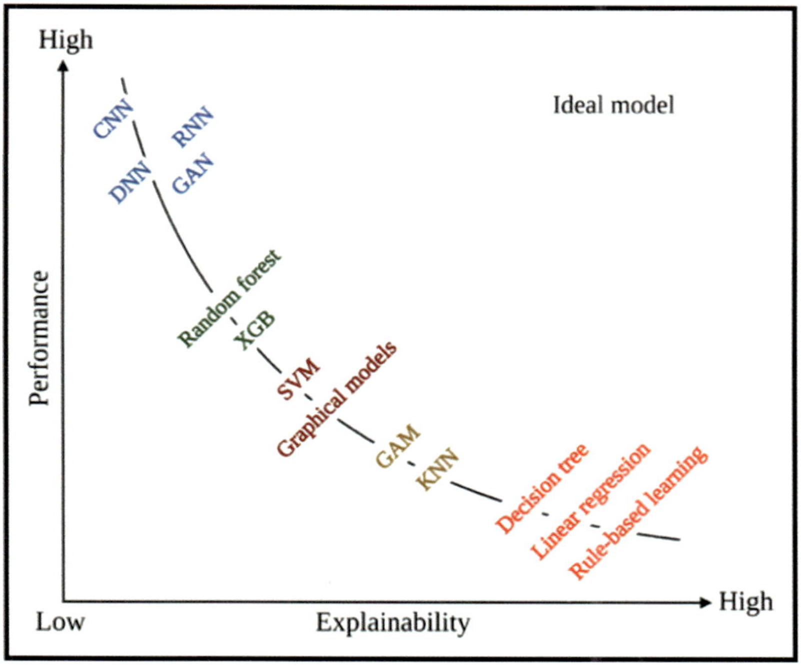

Figure 1: Explainability versus performance for some artificial intelligence methods.

Note: Graphical models: Probabilistic model such as Bayesian network. Rule-based learning: Models using rules (e.g., if: then) for decision-making. CNN: Convolutional neural network; DNN: Deep neural network; GAM: Generalized additive model; GAN: Generative adversarial network; KNN: K-nearest neighbor; RNN: Recurrent neural network; SVM: Support vector machine; and XGB: Extreme gradient boosting. (*Source:* Salih et al., 2023)

To address these issues, researchers and professionals have been developing techniques and approaches to make AI more explainable. This includes using model interpretability methods that help clarify how the model makes decisions based on input data. In this context, Gayane Grigoryan (2022) outlined the following requirements for information to be considered explainable: (i) deliver the content; (ii) provide clarifications; (iii) attempt to justify; and (iv) be faithful to reality.

This chapter will present and discuss recent and relevant studies related to incorporating explainable AI techniques in the health context. The studies are grouped according to their central application, distributed as follows: explainable AI in diagnostic support, explainable AI for procedure personalization, and explainable AI for management. Finally, the narrative review is followed by brief concluding remarks.

Explainable AI in Healthcare

In the healthcare context, AI usage is becoming increasingly prevalent, currently emerging as one of the main tools in the technological transformation of this sector. This is particularly evident in diagnostic support for various pathologies, the provision of personalized treatments, and the optimization of resource and team management. However, factors such as data sensitivity, task complexity, high risks, and the requirement for accountability create unique challenges for AI use in healthcare and justify the need to make its operation transparent and understandable (Chaddad et al., 2023). Transparency can be ensured by detailing how AI algorithms reach their conclusions, which is crucial not only to guarantee the trust of patients and healthcare professionals but also to help identify potential biases and errors in the process (Volkov & Averkin, 2023b).

Therefore, the explainability of the algorithms used becomes an essential feature for their adoption in this sector. It aids in the human interpretation of computational decisions and ensures compliance with applicable regulations. In recent years, numerous studies have focused on this topic, aiming to create and adapt explainable AI techniques for the healthcare context.

Explainable AI in Diagnostic Support

In the study by Tran et al. (2021), a literature review was conducted on deep learning methods applied to cancer diagnostic support. The authors highlight the lack of algorithm explainability as one of the main challenges for their adoption in clinical practice. In the context of medical image analysis, explainability can be incorporated through human validation of clinically relevant regions identified by deep neural networks. In genomics, the authors emphasize the importance of validating relevant genetic features using gold-standard methods.

Volkov and Averkin (2023a) also conducted a literature review on explainable AI methods applied to medical image analysis. Activation or heat maps (CAM and Grad-CAM) are the most common approaches in the studies reviewed. These methods have been used for analyzing X-ray images, MRIs, skin photographs, and images from pathology and histology. The authors note the absence of studies on the usability of explainable techniques in clinical decision support systems and the validation by end users.

Kim et al. (2023) conducted a study to support diabetes control using AI. They tested two hypotheses: (1) whether the comorbidity of diabetes is very strong in middle-aged or elderly people, and (2) whether the determinants of comorbidity are similar for different pairs of diabetes and its comorbid conditions (diabetes-cancer, diabetes-heart disease, and diabetes-mental illness). The authors proposed using the SHAP value (SHapley Additive exPlanations) to quantitatively analyze the association between a determinant and comorbidity. This study supports both hypotheses and highlights the use of SHAP value as an innovation in the field of AI for prediction.

Kang et al. (2023) used deep learning techniques to support autism diagnosis in children based on morphological features in 2,940 facial images. The authors justified the study by the high subjectivity, high cost, and time-consuming nature of the diagnostic process. They acknowledged the responsibility associated with this diagnostic decision-making. Therefore, they applied an explainable AI technique to the evaluated convolutional neural networks (CNNs). They included explainability through a gradient-based analysis method, Grad-CAM (Gradient-weighted Class Activation Mapping), which produces a heatmap highlighting the most heated/active pixels of the image as the most relevant to the final decision. In the study, the weights of the last layer of the CNN were represented as a heatmap, illustrating the areas of the image the neural network focused on to make the diagnostic decision.

Among the CNN architectures analyzed, the ensemble of MobileNet, Xception, and EfficientNet resulted in the best classifications, with an area under the ROC curve of 0.89, a sensitivity of 0.78, and a specificity of 0.82. Despite needing validation by specialized professionals, the study points to the possibility of supporting autism diagnosis through facial image analysis. Moreover, the authors proposed incorporating a technique to illustrate and contribute to understanding the classification process by the artificial neural network, a significant contribution and differential of the work.

In their study, Saravanan et al. (2023) aimed to present a hybrid transfer deep learning model considered more transparent and reliable for supporting the diagnosis of Parkinson's disease (PD) in its early stages. They used wave and spiral drawings from 124 people with PD and 141 healthy individuals. The VGG19 model achieved 98.45% accuracy in classifying people with and without Parkinson's. To make the model more explainable, the authors employed the LIME (Local Interpretable Model-Agnostic Explanations) method to identify parts of the drawings that contributed most to the prediction of the diagnostic class. This study represents a significant contribution to the state of the art in terms of both performance and the incorporation of explainable models. However, further investigations are needed regarding the method's performance on larger datasets and more in-depth interpretations of the highlights made by LIME.

Another diagnostic support study by Baweja et al. (2023) proposed an explainable CNN architecture for predicting leprosy, called LeprosyNet, which achieved 98% accuracy for the task. The study investigated the performances of three explainable AI techniques: Activation Layer Visualization, which generates images of the outputs from each convolutional layer of the neural network;

Occlusion Sensitivity Map, which generates a sensitivity map while part of the image is occluded to assess the effect of occlusion on classification, identifying parts of the image most relevant for decision-making; and Grad-CAM, which highlights parts of the image considered most relevant for classification. Although the study did not provide some relevant information, such as the size of the dataset used, it presented impressive results, with the proposed architecture (LeprosyNet) outperforming well-established models in the literature like AlexNet and ResNet. The application of explainable techniques showed that the CNN model could identify and highlight parts of the image where the lesion existed, demonstrating that these parts were used for decision-making rather than some other nuance of the images. This highlight is crucial for providing greater reliability in the classification process of determining whether a lesion is leprosy or not.

Duraivenkatesh et al. (2023) proposed incorporating SHAP and LIME techniques into pre-trained CNN models like VGG16, VGG19, ResNet50, Inception v3, and Xception to support the early diagnosis of retinoblastoma, a malignant tumor affecting retinal cells. The proposed approach, using transfer learning methods, proved effective in identifying the disease, especially with VGG16, which achieved 98% accuracy. Explainable AI made the models more viable for clinical settings by highlighting the image regions used for decision-making. However, a critical point of the study is the small dataset, with only 280 fundus images, half with the disease and half healthy. This low data quantity also influenced the methodology's choice of transfer learning due to the impracticality of training a robust deep neural network with this image amount.

Studies supporting the diagnosis of Covid-19 also aimed to incorporate explainability into AI-based approaches (Barbosa et al., 2021; Gomes et al., 2023). Gomes et al. (2023), for example, developed a protocol to support disease diagnosis using hematological parameters. They used particle swarm optimization (PSO) combined with human expert interpretation to select tests for diagnosis. Additionally, they developed the system (Heg.IA) using a decision tree-based classification algorithm, graphically explaining its architecture to facilitate the interpretation of the system's final diagnosis of Covid-19. In this context, the adopted explainable AI methodology was crucial not only for healthcare professionals' adherence to the system but also for understanding the disease itself and its effects on hematological parameters, which were not well-explored before this study.

Explainable AI for Personalization of Procedures

The adoption of explainable AI techniques has also contributed to personalizing healthcare, making it more suitable for the individual needs and characteristics of patients. In this context, the study by Nazar et al. (2021) conducted a systematic literature review on the intersection of human-machine interfaces and explainable AI in healthcare. Through a comprehensive survey of the state-of-the-art, the authors reiterated the importance of explainable methods to improve user trust, involving people in the care process. They also highlighted several current challenges related to the incorporation of AI in healthcare, such as the adoption of black-box models, distrust in human-AI communication, resistance to adopting AI-based systems, auditing AI output, computational costs, and the need for significant infrastructure changes. Many of these challenges, they noted, could be mitigated by adopting explainable methods.

In addition to these challenges, Paranjape et al. (2020) addressed the need to include AI training in the education of healthcare professionals as a way to familiarize them with the techniques and their understanding. Another important point brought up by the authors is the role of explainable AI methods in simplifying the regulatory process for AI use by competent authorities in each country, a topic also discussed by O'Sullivan et al. (2019) regarding the use of explainable AI to support accountability in using autonomous surgical robots.

Furthermore, Paranjape et al. (2020) mentioned how various AI-based techniques have already been developed and show high performance in solving complex healthcare problems. These techniques can be combined to provide more in-depth analyses of a patient's overall health condition. As an example, the authors cited techniques in the context of lung cancer, where machine learning algorithms have been trained to enhance understanding of critical genomic pathways, predict which patients will respond to disease inhibitors, and personalize radiotherapy, thereby choosing an optimal treatment strategy.

This broad and combined analysis of various aspects related to a health condition is also discussed in the study by Lipkova et al. (2022). The authors explored the use of AI for integrating multimodal data in oncology. In this context, the patient's condition is characterized by various modalities, ranging from radiology, histology, and genomics to electronic health records, each providing additional information. Using these data together through multimodal models is crucial for a more comprehensive understanding of the problem. The authors argue that explainable models can play a significant role in exploring the contribution and importance of each modality in the observed outcome. This contributes to clinical analysis and managerial issues related to screening, diagnosis, and treatment processes in a more personalized manner.

Explainable AI for Management

The quality of healthcare heavily relies on managerial and administrative factors, where algorithms and intelligent systems can contribute significantly through automation and prediction. Yang (2022) exemplifies the use of predictive models to support monitoring patient progress and improve documentation and records in general. The author states that computational advances, especially cloud computing, big data, and the internet of things, have enabled the connection of disparate data and devices and large-scale storage, facilitating the development of predictive health models. However, the current models are often unintuitive and difficult to explain. Enhancing explainability is essential to promote the adoption and utility of these models. It allows for understanding and addressing questions such as 'why', 'why not', 'what if', 'where', 'when', 'how could this be', and 'how still could this be' related to a specific scenario.

Examples of explainable AI models for prediction were developed in the studies by Alabdulhafith et al. (2023) and Lauritsen et al. (2020). In the first study, the authors proposed a machine learning model to predict readmission in the Intensive Care Unit (ICU). Models like this are crucial for reducing risks and mortality rates by providing information that prevents improper patient discharge, promotes appropriate interventions, and aids in planning patient allocation post-ICU discharge. The model was based on analyzing vital signs, laboratory tests, and patient demographic data. A distinguishing feature of the study was the emphasis on providing local and global explanations of the best-performing model using SHAP and LIME techniques to verify its clinical validity and generalizability.

The SHAP technique was also applied in the predictive model proposed by Lauritsen et al. (2020). They used a variation of the CNN called temporal convolutional network (TCN) to analyze electronic health record data for predicting acute critical illnesses (e.g., sepsis, acute kidney injury, and acute lung injury). They proposed an explainable intelligent early warning score system (XAI-EWS) for detecting these illnesses in very early stages, achieving an area under the ROC curve of around 0.8.

Patient electronic health records (EHRs), as used in the studies mentioned earlier, are essential tools for managing and tracking the patient's journey within and outside a healthcare facility. They connect various types of information related to the individual and their health condition. However, EHR data remain underexplored due to data protection issues, as most of them are personal and sensitive, and challenges related to the automatic analysis of some free-text fields filled out by healthcare professionals. To address the latter challenge, approaches like the one adopted in the study by Górski and Ramakrishna (2021) can be utilized. The authors proposed applying a CNN for text classification. They added explainability to the model by highlighting parts of the text that were most important for decision-making, using Grad-CAM, SHAP, and LIME models. In the study, they compared classification performance and explainability analysis based on accuracy and legal validation of the explanation by human experts.

From a management perspective, explainable AI models can also be useful in clinical engineering and technovigilance. Techniques for machine integrity prognosis, like the one proposed in Hong et al. (2020) study, which is based on long short-term memory (LSTM) and SHAP, can support the management and maintenance of medical equipment. Additionally, managing contracts, analyzing the acquisition of health products and equipment, and prioritizing patient care in healthcare facilities can be improved with explainable AI models. For instance, Han and Lee (2021) proposed a model based on deep neural networks for these purposes.

Final Considerations

The incorporation of XAI techniques in healthcare represents a significant advancement in ensuring transparency, ethics, and efficacy of AI systems. By providing understandable explanations of algorithmic decisions, XAI fosters greater trust among healthcare professionals, managers, and patients. This trust is crucial for making critical decisions that directly impact the lives of individuals within this ecosystem.

The studies presented in this chapter showcase the diverse applications of XAI in healthcare, ranging from diagnostic support to personalized procedures and health management. Researchers have employed model interpretability methods and innovative approaches to address the challenges posed by the lack of transparency and explainability in AI-based systems.

However, the continuous development of XAI techniques is essential to meet the ever-evolving challenges in healthcare, especially those related to data protection and decision auditing. Collaboration among researchers, healthcare professionals, regulators, and patients is vital to validate AI-based solutions in terms of utility, usability, ethics, safety, and efficacy (Volkov & Averkin, 2023b).

In the healthcare domain, XAI not only enhances the acceptance and adoption of technology but also strengthens its utility and efficacy in delivering patient-centered, high-quality care while mitigating legal and ethical risks associated with the use of AI algorithms in clinical settings.

References

Alabdulhafith, M., Saleh, H., Elmannai, H., Ali, Z.H., El-Sappagh, S., Hu, J.W., & El-Rashidy, N. (2023). A clinical decision support system for edge/cloud ICU readmission model based on particle swarm optimization, ensemble machine learning, and explainable artificial intelligence. *IEEE Access*, 11, 100604–100621.

Barbosa, V.A.F., Gomes, J.C., Santana, M.A., Albuquerque, J.E.D.A., Souza, R.G., Souza, R.E., & Santos, W.P. (2021). Heg. IA: An intelligent system to support diagnosis of Covid-19 based on blood tests. *Research on Biomedical Engineering*, 1–18.

Baweja, A.K., Aditya, S., & Kanchana, M. (2023). Leprosy diagnosis using explainable artificial intelligence techniques. *In*: *2023 International Conference on Sustainable Computing and Data Communication Systems* (*ICSCDS*) (pp. 551–556). IEEE.

Chaddad, A., Lu, Q., Li, J., Katib, Y., Kateb, R., Tanougast, C., Bouridane, A. & Abdulkadir, A. (2023). Explainable, domain-adaptive, and federated artificial intelligence in medicine. *IEEE/CAA Journal of Automatica Sinica*, 10(4), 859–876.

Duraivenkatesh, S., Narayan, A., Srikanth, V., & Made, A.F. (2023). Retinoblastoma detection via image processing and interpretable artificial intelligence techniques. *In*: *2023 IEEE Conference on Artificial Intelligence* (*CAI*) (pp. 166–167). IEEE.

Gomes, J.C., Freitas Barbosa, V.A., Santana, M.A., de Lima, C.L., Calado, R.B., Bertoldo-Júnior, C.RB., Albuquerque, J.E.A., Souza, R.G., Araújo, R.J.E., Moreno, G.M.M.,

Soares, L.A.L., Mattos-Júnior, L.A.R., Souza, R.E. & dos Santos, W.P. (2023). Rapid protocols to support Covid-19 clinical diagnosis based on hematological parameters. *Research on Biomedical Engineering*, 39(3), 509–539.

Górski, Ł., & Ramakrishna, S. (2021). Explainable artificial intelligence, lawyer's perspective. *In*: *Proceedings of the Eighteenth International Conference on Artificial Intelligence and Law* (pp. 60–68). ACM: USA.

Grigoryan, G. (2022). Explainable artificial intelligence: Requirements for explainability. *In*: *Proceedings of the 2022 ACM SIGSIM Conference on Principles of Advanced Discrete Simulation* (pp. 27–28). ACM: USA.

Han, J., & Lee, Y. (2021). Explainable artificial intelligence-based competitive factor identification. *ACM Transactions on Knowledge Discovery from Data* (*TKDD*), 16(1), 1–11.

Hong, C.W., Lee, C., Lee, K., Ko, M.S., & Hur, K. (2020). Explainable artificial intelligence for the remaining useful life prognosis of the turbofan engines. *In*: *2020 3rd IEEE International Conference on Knowledge Innovation and Invention (ICKII)* (pp. 144–147). IEEE.

Kang, H., Yang, M., Kim, G.H., Lee, T.S., & Park, S. (2023). DeepASD: Facial image analysis for autism spectrum diagnosis via explainable artificial intelligence. *In*: *2023 Fourteenth International Conference on Ubiquitous and Future Networks* (*ICUFN*) (pp. 625–630). IEEE.

Kim, R., Kim, C.W., Park, H., & Lee, K.S. (2023). Explainable artificial intelligence on life satisfaction, diabetes mellitus and its comorbid condition. *Scientific Reports*, 13(1), 11651.

Lauritsen, S.M., Kristensen, M., Olsen, M.V., Larsen, M.S., Lauritsen, K.M., Jørgensen, M.J., Lange, J., & Thiesson, B. (2020). Explainable artificial intelligence model to predict acute critical illness from electronic health records. *Nature Communications*, 11(1), 3852.

Lipkova, J., Chen, R.J., Chen, B., Lu, M.Y., Barbieri, M., Shao, D., Vaidya, A.J., Chen, C., Zhuang, L., Williamson, D.F., Shaban, M., Chen, T.Y., & Mahmood, F. (2022). Artificial intelligence for multimodal data integration in oncology. *Cancer Cell*, 40(10), 1095–1110.

Nazar, M., Alam, M.M., Yafi, E., & Su'ud, M.M. (2021). A systematic review of human–computer interaction and explainable artificial intelligence in healthcare with artificial intelligence techniques. *IEEE Access*, 9, 153316–153348.

O'Sullivan, S., Nevejans, N., Allen, C., Blyth, A., Leonard, S., Pagallo, U., Holzinger, K., Holzinger, A., Sajid, M.I., & Ashrafian, H. (2019). Legal, regulatory, and ethical frameworks for development of standards in artificial intelligence (AI) and autonomous robotic surgery. *The International Journal of Medical Robotics and Computer Assisted Surgery*, 15(1), e1968.

Paranjape, K., Schinkel, M., & Nanayakkara, P. (2020). Short keynote paper: Mainstreaming personalized healthcare–transforming healthcare through new era of artificial intelligence. *IEEE Journal of Biomedical and Health Informatics*, 24(7), 1860–1863.

Salih, A., Boscolo Galazzo, I., Gkontra, P., Lee, A.M., Lekadir, K., Raisi-Estabragh, Z., & Petersen, S.E. (2023). Explainable artificial intelligence and cardiac imaging: Toward more interpretable models. *Circulation: Cardiovascular Imaging*, 16(4), e014519.

Saravanan, S., Ramkumar, K., Narasimhan, K., Subramaniyaswamy, V., Kotecha, K., & Abraham, A. (2023). Explainable Artificial Intelligence (EXAI) models for early prediction of Parkinson's disease based on spiral and wave drawings. *IEEE Access*, 99, 1–1.

Tran, K.A., Kondrashova, O., Bradley, A., Williams, E.D., Pearson, J.V., & Waddell, N. (2021). Deep learning in cancer diagnosis, prognosis, and treatment selection. *Genome Medicine*, 13, 1–17.

Volkov, E.N., & Averkin, A.N. (2023a). Explainable artificial intelligence in medical image analysis: State of the art and prospects. *In*: *2023 XXVI International Conference on Soft Computing and Measurements* (*SCM*) (pp. 134–137). IEEE.

Volkov, E.N., & Averkin, A.N. (2023b). Explainable artificial intelligence in clinical decision support systems. *In*: *2023 IV International Conference on Neural Networks and Neurotechnologies* (*NeuroNT*) (pp. 3–6). IEEE.

Yang, C.C. (2022). Explainable artificial intelligence for predictive modeling in healthcare. *Journal of Healthcare Informatics Research*, 6(2), 228–239.

9 | Longevity and Exclusion: Ethical Challenges Related to the Use of Artificial Intelligence in Elderly Care with Dependency

Ciro Augusto Floriani* and Hernán Poblete Miranda*

This chapter analyzes, based on the report of some successful experiences, the potential risks and harms in the development and applicability of Artificial Intelligence (AI), Machine Learning, and Robotics in the care of dependent elderly individuals, including those in palliative care. It particularly examines the operational difficulties and the logic behind the construction of these tools in a society like ours, notably marked by structural, intergenerational inequality and ageism. The chapter emphasizes the importance of including bioethicists in the construction of these decision-making and intervention models to ensure that the protection of this vulnerated population is the central reference in this new and expanding system of surveillance and care. This will have impacts on the quality of life and the quality of death for elderly individuals.

Getting old is exactly like having a criminal record.
From now on, every little thing counts against you.

— Marie Baird

Introduction

Brazil is experiencing a demographic transition characterized by increased longevity among the elderly population. This transition involves reduced fertility rates and a persistent decline in mortality, leading to a rise in non-communicable diseases in this age group, with high prevalence of conditions like cancer and dementia (Veras, 2002).

*Corresponding authors: ciroafloriani@gmail.com; hernanpoblete@gmail.com

One consequence of this scenario concerns the quality of life for the elderly and the allocation of resources needed to organize and structure healthcare services for this growing population segment. This presents a challenge for health policies in both the public and private sectors, as suitable and highly effective care models become increasingly necessary in a context of budgetary constraints.

The complexity intensifies with the scarcity of healthcare resources, stagnant and frozen budget caps, and rising costs. There are undeniable setbacks in the already precarious social welfare network, resulting from the recent neoliberal policies legislated in our country and implemented in recent years, particularly since 2016. Reconstruction is imperative.

In this essential reconstruction, it is crucial to consider that the growing elderly population has been eroded by labor and social security laws that have stripped away historical rights. The productive demands imposed on this population are cruel, exacerbated by the social inequality process, the inherent limitations of their longevity, and associated health issues. The situation is difficult and complex, and simplistic solutions will not resolve it. Distributive justice faces a long-standing battle that has intensified in recent years, in an environment where an economic reductionism of the healthcare reality prevails, resembling an unjustifiable 'tyranny' of one sphere of justice over others (Waltzer, 1983).

In this context, biotechnology emerges as an alternative solution to the complex problem of healthcare access for millions of people (Schramm, 2010). Specifically, Artificial Intelligence (AI) and its ongoing applications in health become fundamental actors in addressing the issues of exclusion and marginalization faced by the growing population without access to quality healthcare.

As detailed in this chapter, AI can facilitate patient communication and provide emotional support through apps and teleassistance programs, fostering closer connections with loved ones and healthcare professionals. It can also enhance the speed and accuracy of diagnoses and disease detection, simplifying medical care. Additionally, AI can bolster health research, drug development, and various public health initiatives, while giving patients greater control over their medical care and a deeper understanding of their evolving needs[1].

However, to what extent is this promise of integration real? Does it conflict with privacy protection rights or basic non-discrimination norms, which are often accepted as inevitable externalities? Though these issues are currently less visible, they could lead to increasing conflicts in healthcare access in the near future.

However, to what extent is this promise of integration real, and does not it conflict with privacy protection rights or basic non-discrimination norms? These measures are often accepted as inevitable externalities, and although they are not yet highly visible, they could lead to increasing conflicts in healthcare access in the near future.

---

1 Authors' note: Citation from BBC News Portuguese. (2019). Como a Inteligência Artificial está revolucionando o diagnóstico médico. https://www.bbc.com/portuguese/vert-fut-47296297.

This chapter will provide examples of the applicability of AI and Machine Learning (ML) for elderly individuals with increasing dependency, as well as those with advanced and terminal illnesses. Through these experiences, we will analyze the risks of applying these devices to dependent elderly care. We are particularly interested in the operational difficulties and the underlying logic of AI and ML development, as well as their potential risks and harms. These issues arise especially when the construction of these decision and intervention models does not consider the importance of Bioethics tools (Lewis, 2020).

There is a need for a protective perspective in building a care model during a period of increased physiological limitations. Health issues can further reduce the quality of life, leading to unpredictable and undeniable impacts on the elderly's daily lives. This results in the progressive loss of independence and, depending on the severity, also impacts their autonomy. Ultimately, these models will be applied to lives full of restrictions. The variables used in their construction could further deteriorate the quality of life and death for dependent elderly individuals.

The central question for us is whether humans serve AI or if AI serves humans. Do predictive models used as devices amplify the essence of care in healthcare, or do they amplify the abandonment to which a significant portion of elderly individuals in our country is already subjected? We understand that some risks involve enclosing humans in cyber spaces that feed on *ad hoc* decision-making processes, which do not consider relevant moral aspects such as protection, non-abandonment, and beneficence. Thus, we must ask: Do those who design AI systems consider the needs and desires of the elderly? The impact of these models will undoubtedly be significant, as this technology will not only become more accessible in "hospital corridors" but increasingly so at our "fingertips and eyes" in our private lives.

Longevity, Dependency, and Intergenerational Enclaves

All of this would merely be information if it weren't for the revolution in longevity. This revolution represents a systemic change in the social structural order caused by an unprecedented increase in life expectancy. This variation exacerbates the already unequal living conditions of millions, who now face long lives under circumstances of economic impoverishment that push them towards marginalization. The ethical conflicts arising in this context of worsening life conditions underscore the need to address critical questions, such as the necessity of allocating more resources to elderly healthcare. Additionally, there is a need to determine the burden of these allocations and who should bear it, especially in relation to the resources used for younger generations, whose living conditions are also becoming increasingly precarious.

Moreover, the social ecosystem that gives meaning to the lives of the elderly in today's globalized urban spaces is determined by the individual's socioeconomic and functional status. Their fate seems to lead towards marginalization based on their ability or inability to remain productive. The distinction between "mobility/access" (especially in urban contexts) will play a crucial role in their integration into or exclusion from the ecosystem once young (functional) individuals become 'elderly'. Old age and extreme dependency will become the ultimate defeat, marked by their expulsion from the productive system that had given their lives meaning for most of their existence (Poblete, 2018).

In a system where competitive intergenerational and individualistic productivity prevails, removing the elderly from the productive space also means removing them from the public space and restricting their freedom of movement. As their capacities and functional reserves diminish and their vulnerability to falls and harassment increases, elderly individuals gradually lose the ability to navigate essential aspects of daily life. These include public transportation, walking on neighborhood sidewalks, or completing tasks at a bank. Streets, pedestrian bridges, and even information systems become critical and indecipherable, suitable only for those with adequate vision and hearing capabilities due to their age. Urban spaces are designed and maintained for the productive functionality of the young. Those who cannot keep up are expected to withdraw.

If the elderly resist, perhaps by trying to improve their economic situation by working beyond retirement age because their meager pensions are insufficient, other passengers on public transportation will often make it clear that this is not their place. The elderly may be told they don't belong as they delay the bus with their slow boarding and alighting or hold up the supermarket line during peak times when young workers have only a brief lunch break before returning to work.

In other words,

> "The elderly are aware of this distinction that marginalizes them because they experienced it before becoming elderly. They were young once and viewed the elderly of their time similarly. Therefore, they know they are seen as a hindrance to the productive system's flow, an immovable obstacle to economic growth that is not tolerated. They sense (or believe they sense) the general indifference to their condition or stage of life, but they find solace in the thought that many of those who mistreat them are unaware that they will face old age sooner or later in their own lives or in the life of a close relative" (Poblete, 2018, p. 58).

It is precisely the abandonment and humiliation that the elderly suffer when exposed to all social systems, including medical-sanitary systems, that is the consequence of being seen as obsolete objects. This is the crux of their victimization in an intergenerational war caused by the struggle for survival in the ruthless and insensitive context of productive conflict. In other words, this inequality in healthcare originates from the tyranny of a society that allocates the role of bodies in the productive landscape (Citro, 2010). If one is young, one is effective and, therefore, useful. If one is old, one is unproductive and useless, and thus this body must make way for fresher, more vital bodies in productive spaces. This scenario no longer describes a class struggle to control the means of production but rather a cruel battle between generations, where each age group pushes and marginalizes the other from urban and productive spaces, relegating them to exile and solitude until death (Poblete, 2018).

Considering longevity as an interconnected ecosystem, where symbolic dimensions impact cultural aspects, and viewing the elderly as those requiring specific and costly care related to all elements of the social system but "without giving anything back to society", the "without giving back" means: "whatever they did, however significant, is in the past and now is history" (Poblete, 2018, p. 58). It is as if society does not see them as its problem but as a necessary evil present in the picture of the landscape: younger, considered productive, individuals will see the elderly as a problem they have to endure. Thus, the tyranny of health inequality can be perceived as the elderly's problem: it will be their responsibility to address it. Being old will then become an individual problematic condition, not even considered in any aspect as a positive experience collectively or individually (Thomas, 1983).

This constant segregation of the elderly by the younger population does not imply that the current productive protagonists, such as urban bus drivers, display aggressive behavior exclusively towards the elderly.

On the contrary, the elderly appear to be victims of the general assumption that they are a problem that needs to be endured and limited. Thus, the bus driver himself becomes one of these limits. Anointed as the main actor and empowered to mediate access to public transportation, he carries the exclusion and denial of rights to this group of people who, as it is often said, "are useless" or "hinder the lives of the active". If the driver has become the actor in this case, it is because he has been chosen for the terrible task of ending the elderly urban dweller's ability to move, much like an executioner hired and paid to permanently end the lives of the condemned. Ultimately, it is the driver who imposes the final limit on the elderly person's possibility (and human right) to continue moving through public space and thriving. From there, from this insurmountable limit for this being tormented by the fear of falling and breaking, will begin his gradual reclusion and the abandonment of his desire to remain functional and productive. This marks the beginning of his social death (Poblete, 2018).

> "One can consider that there is a social death (with or without actual biological death) whenever a person ceases to belong to a given group, whether due to age limits and loss of functions (deceased and dead are related), or by being subjected to acts of degradation, proscription, exile, or even when we face a process of erasing memory (disappearance without leaving traces, at least at the level of consciousness)" (Thomas, 1983, p. 68).

Social death emerges in our lives when we reach these age limits that longevity has opened. It is possibly worse than biological death, a living burial for those who once worked and were active producers of wealth but now, in old age, "serve no purpose". Social death means losing one of the essential dimensions of human existence, one's social life, which defines them as a subject in biological existence while extending their life. The issue of age teaches roles and social representations where a set of values marks the differences and expectations of these roles.

> "Passing through a time when one experiences fulfilling these roles until a time when none remain, not even the slightest expectation. Revealing one's biological age can lead to discrimination precisely because of what age can signify socially. In his dual existence, he increasingly feels a contrast between the ever more meaningless social being he carries in body and mind and his biological reserve of life. The day he feels that only this biological reserve still has the right to exist, without any external purpose, he will feel, consciously or not, the arrival of social death" (Thomas, 1983, p. 68).

Here, there is an intergenerational enclave that places at the center of the debate the question of whether distributive justice has a privileged space as a key ordering principle in the construction of AI and ML decision-making processes. The extreme culture of productivity excludes the elderly and deepens the intergenerational conflict, exacerbating the inequality of access that becomes a permanent reality for all generations (Callahan, 1995).

Within this scenario of increasing elderly dependency, although the process of dying can be viewed from a more concrete, profound, and reflective perspective—considering the environment and the living presence of the social ecosystem that gives it meaning and recognizing it as a complex and multifaceted issue involving emotional, cultural, social, and spiritual aspects—the reality seems different: the end of the window of demographic bonus[2] indicates that millions of people are joining the ranks of the elderly with comorbidities and resource demands. This situation forces the system to choose between allocating increasing economic resources to productive youth or to an army of "passive people" who consume these same resources for longer periods.

---

[2] Authors' notes: The conclusion of the demographic bonus in many countries is presenting significant implications and challenges in healthcare, pensions, and other social support systems, pressured by the increasing proportion of elderly individuals (WHO, 2015). The demographic bonus refers to a period when a population has more people of working age than dependents (typically those under 15 and over 64). This dynamic can provide an economic boost due to a larger workforce compared to the number of dependents, offering an opportunity for economic growth and development if leveraged effectively (UNFPA, 2014). The idea behind the demographic bonus is that it can lead to increased economic productivity, higher savings rates, and even spur innovation and entrepreneurship. However, this 'bonus' is temporary and requires effective management. To fully capitalize on the demographic dividend, a nation needs to invest in education and training, promote an inclusive and diverse economy, and ensure appropriate labor and tax policies. If these conditions are not met, the country

(Contd.)

Excluding the elderly from participating in the construction of these algorithmic systems, without considering their needs and desires, sustains a stereotypical view of what it means to be old. It reinforces a hypertrophied societal view of the elderly as passive beings, without desires or autonomy, molded by *ad hoc* decisions about their needs, turning them into a product rather than subjects of their time (WHO, 2022). Therefore, alerts about the risks and suggestions on the importance of elderly protagonism in the construction of these predictive AI models are very welcome (Mannheim et al., 2019).

Some Experiences of Artificial Intelligence and Machine Learning with the Elderly

Infused with optimism, gerontechnology, understood as the use of AI and ML tools for the elderly, is a reality here to stay, expanding and improving worldwide (Chen, 2020). Physical and recreational activities at home, approaches for frail elderly, those with dementia, and end-of-life care are fields where various experiences have shown success, including robotic assistance, although these need further temporal consolidation (Chen et al., 2020; Dolatabadi et al., 2019; Huang et al., 2019; Liu et al., 2020).

Robotic therapy, an innovative form of therapy using robots, has proven effective in supporting elderly individuals in various settings, including daycare centers and healthcare institutions. For example, Pepper, the world's first humanoid robot capable of identifying faces and primary human emotions, has been used at the AZ Damiaan Hospital in Ostend, Belgium. It assists patients with reduced mobility to remain active and socially connected, despite its technical limitations[3], by helping them socialize and participate in recreational activities.

---

may face significant challenges when the proportion of the working-age population begins to decline, and the number of dependents rises, known as the "burden of aging" (Lee & Mason, 2011). The end of the demographic dividend is not occurring universally across the globe at the same time. The phenomenon occurs at different times for different countries, depending on their demographic evolution. In developed countries such as Western Europe, Japan, and the United States, the demographic bonus is advanced or has already ended. In these countries, population longevity results in a higher proportion of dependents (those over 65) compared to the working-age population (United Nations, 2019). In developing countries, the situation varies. For instance, in Africa and parts of Asia, a continuous demographic dividend exists due to a young population. However, this bonus may not be fully utilized due to factors such as a lack of employment and education opportunities for the youth. Conversely, countries like China and India, which have experienced a significant demographic bonus in recent decades, are beginning to see this bonus diminish due to declining birth rates and the gradual aging of their populations (Bloom et al., 2011) (See References).

[3]Author's notes: Clusters&Ie. (2023) (See References).

In Japan, the therapeutic robot PARO is even more advanced and differs from Pepper with its innovative design used since 2003. Widely utilized to help reduce stress and anxiety in elderly patients, PARO remains under constant research to assess its impact. It has five types of sensors, including touch, light, hearing, temperature, and posture sensors, allowing it to perceive people and its environment. A sensitive photosensor helps PARO recognize light and dark, while the touch sensor allows it to feel when it is petted or tapped. The direction of voice and words like its name, greetings, and praise, thanks to its audio sensor, or its posture sensor, make it even more real to the patient, fostering an empathetic relationship between human and machine. Its goal is to be a "therapeutic robot", capable of learning to behave in the manner preferred by the user and responding to its new name. If a patient pets PARO each time they touch it, the robot will remember the previous action and try to repeat it to be petted again[4].

To evaluate the effects of Paro on the elderly, studies have been conducted since 2009 involving the measurement of mood using facial scales and questionnaires. Changes in their stress responses were also examined by measuring urinary hormone levels. The results of these studies were remarkable. During the period when the elderly interacted with Paro, a significant improvement in mood and stress-coping ability was observed, as evidenced by the urine samples collected. Additionally, the stress levels of the caregiving staff decreased because the elderly required less supervision during interaction with the robot. Continuous improvements in the feelings of the elderly and a decrease in depression levels were noted. Caregivers observed that the interaction had a positive effect on the happiness and activity levels of the elderly, and their facial expressions became softer and more joyful. Participants eagerly anticipated the interaction sessions with Paro, creating a more positive and engaging environment. Communication between the elderly and their caregivers also improved, with the interaction with the robot becoming a common topic of conversation, enhancing the social dynamics (or resulting in improved vital organ responses to stress, as evidenced by urine analyses)[5].

The use of innovative and advanced tools for reducing stress and anxiety, as well as for the care and well-being of elderly patients, is already a reality, and PARO is no exception. The assistive robot Zora is widely used in nursing homes and healthcare units in over 30 countries and has become a widely utilized tool for patient care and well-being. Success stories of Zora include its use at Jouarre Hospital in Paris, where it has been used to help patients with daily activities such as standing up and sitting down, exercising, playing, taking medication, and remembering medical appointments. In the United States, Asia, and the Middle East, it has been used to stimulate physical and cognitive activity in elderly patients with neurological diseases and dementia. Additionally, it has been used

---

4 Authors' notes: Clusters&Ie. (2023). Paro Robots. (2023) (See References).

5 Authors' notes: Therapeutic Paro robots help improve the quality of life for people with dementia and other conditions (Openmind, 2017) (See References).

in nursing homes to assist patients with daily tasks such as reading and making phone calls, helping patients stay connected with their loved ones[6].

During the Covid-19 pandemic, some robots also played a crucial role in supporting healthcare professionals before patient admission. A notable case is in Mexico, where a startup called Roomie developed an autonomous robot that provided valuable assistance to medical personnel caring for high-risk Covid-19 patients. This innovative robot, named RoomieBot, played an essential role in triaging patients by assessing their temperature, blood oxygen level, and medical history before hospital entry. RoomieBot uses advanced technology based on Intel, including artificial intelligence algorithms executed on its vision processing unit, to perform these tasks efficiently and accurately[7].

At Hospital Padre Hurtado in Chile, the first robot designed to support patient care during the Covid-19 pandemic was introduced. Named 'Eva', this robot was installed through a private initiative and aimed to serve as the initial point of contact between patients and the medical team. Patients can interact with 'Eva' via a screen to communicate their needs, describe their condition, and receive responses or be directed to the appropriate care. Eva is equipped with a range of sensors, a battery with eight to twelve hours of autonomy and automatic recharging capability, a camera for patient visualization and recognition, a microphone, internet connectivity, full 360-degree mobility, and an autonomous intelligent navigation system. Its purpose is to enhance hospital procedure safety, promote social distancing, and prevent potential exposure of medical staff to the virus[8].

Despite the understandable enthusiasm for gerontechnology, it is crucial to remain attentive to the risks of biases in developing these cyber tools. In robotics, for example, there are cases where relationships are configured to appear 'warm' and 'affectionate'. We are familiar with such relationships among living beings, but it is something different and new between a human and a robot. Even if a robot can be programmed to be affectionate and warm, the absence of human interaction with the elderly cannot be overlooked. Have we individualized ourselves to the point where we can do without human presence? Or are we witnessing a process of separation, exclusion, and isolation becoming naturalized?

The man filled with himself, the homo clausus of Elias (2001), characterized by loneliness, disconnection, and isolation, stems from a conception where "each one exists only for himself, independent of other human beings and the entire 'external world'... and with it the idea that a person must have a sense exclusively their own" (p. 65). This complex process of individualism is well captured in the words of the Portuguese poet Fernando Pessoa, who says, "a deserted street

[6] Authors' notes: In France, a robot named Zora helps the elderly stay active and socially engaged (NYT, 2018) (See References).

[7] Authors' notes: Intel describes how robots are transforming healthcare by assisting with tasks such as transportation, disinfection, and social interaction (Intel, n.d., 2023) (See References).

[8] Authors' notes: The Hospital Padre Hurtado in Chile implemented the first robot to support patient care during the Covid-19 pandemic (Red Salud, 2020) (See References).

is not a street where no one passes, but a street where those who pass, pass as if it were deserted" (Fernando Pessoa, 1982, p. 67). Today, we might say they pass as if reality is in their cell phones and not in encounters on the street. If it is true that the robot meets the physical needs and daily tasks of the dependent elderly, and supposedly provides some affection, it is equally true that this robot cannot replace the human being in the encounter with others.

And encountering the other is what makes it possible to develop the conditions for an "ethics of responsibility", within a conception where

> "ethics is a relationship with others, with the neighbor... **the contingent nature** [author's emphasis] of this relationship, ... I can only emerge in my uniqueness by responding to others through a responsibility from which there is no possible escape, a responsibility I can never fully discharge" (Dufourmantelle, 2003, pp. 45–46).

From this point on, it is no longer about me, but about the other within me, and "this responsibility goes as far as fission, to the denuclearization of the self. And therein lies the subjectivity of the self" (Levinas, 2003, p. 152). Are we truly providing the elderly with what they essentially need, or, within a stereotyped view of their needs, are we creating an environment of illusions that perpetuate abandonment and isolation?

Artificial Intelligence, Machine Learning, and the Good Death

Another potential application of AI lies in end-of-life care. AI and ML could play a significant role in improving the dying process. Due to their widespread use and potentially lower financial costs, they may help optimize referrals to Palliative Care, potentially enhancing the quality of death (Avati et al., 2018; Murphree et al., 2021; Santos et al., 2020).

AI has been used to help monitor and control symptoms in palliative care (Shimada & Tsuneto, 2023). It can also assist in developing the Advanced Care Plan, aiding in complex decision-making, reviewing treatments, reducing medication-related iatrogenesis, and identifying patient preferences and values (Medinaceli & Silva, 2021)[9]. Additionally, some AI systems can collect and

---

[9] Authors' notes: Artificial Intelligence (AI) has been used to assist in treatment review and reduce medication-related iatrogenesis—adverse effects resulting from medical treatment. AI applications in healthcare can integrate expert-approved algorithms with electronic health record data to review different treatment alternatives and recommend the most suitable combination for each patient. By providing a comprehensive and precise review of treatment options and personalizing therapeutic recommendations for each patient, AI can help minimize the risk of adverse effects and improve treatment efficacy.

analyze large amounts of patient health data to predict life expectancy with some accuracy. For example, studies are underway to predict the survival of hospitalized patients (Tomašev & Glorot, 2019)[10]. This capability can help doctors and patients engage in informed conversations about advanced care planning. Decisions on when to withdraw medical treatments can be assisted by AI tools, which can use data from previous patients with similar conditions and treatments to inform these decisions. Similarly, AI and Natural Language Processing (NLP) can analyze conversations or written texts to identify an individual's values and preferences, which is useful for understanding how a patient may want to be cared for in their final days.

AI has been understood as playing a significant role in elderly healthcare, with its expanding capacity showing great promise. However, it's crucial to approach its use ethically and responsibly, always prioritizing the patient's benefit and respecting their rights (WHO, 2021a). Lack of regulation can lead to age-based discrimination and undermine the quality of care provided to this population. The global health agency warns of these risks and proposes policies to mitigate them (Malpani et al., 2022).

Regarding connections with loved ones, there are AI tools that enable interaction with those who have passed away. For example, HereAfter.AI[11] is an interactive memory application that preserves stories and memories, allowing users to interview themselves about their lives and record these stories in audio. Loved ones can listen to the stories and view photos through the app. It includes hundreds of inspiring questions to help share memories, facilitating conversations with deceased loved ones.

Another example is Rememory, a platform developed by the South Korean company DeepBrain AI, which allows people to connect with their deceased loved ones virtually, supported by artificial intelligence. Rememory collects digital memories of the deceased, such as videos and photographs, and uses this information along with a prior interview with a family member to organize a virtual meeting. The tool offers two options: receiving a pre-recorded message from the deceased or having a real-time conversation with them. To facilitate the meeting, a special room equipped with a 400-inch screen and high-quality sound systems is required to make the conversation as realistic as possible. This application was recognized in the Virtual and Augmented Reality category at the 2023 Consumer Electronics Show Innovation Awards[12]. These systems rely on natural language processing AI to recreate human-like interactions.

[10] Authors' notes: The authors demonstrate the use of AI to predict acute kidney injury in hospitalized patients, which can be relevant in medical decision-making, but it is not specifically for predicting hospital mortality.

[11] Authors' notes: HereAfter.AI is a company that creates digital avatars powered by artificial intelligence to preserve the memories and personalities of loved ones (Hereafter. AI, n.d.) (See References).

[12] Authors' notes: Rememory is a platform that uses artificial intelligence to create digital avatars of deceased people, allowing users to interact with them (Entrepreneur, 2023) (See References).

Nevertheless, being a field still under study and expansion, it's crucial not to lose sight of the true purpose of end-of-life care, which is to help the patient achieve a good death by strengthening human interactions (not with robots) (Nwosu et al., 2019). This involves improving diagnostic capabilities for appropriate palliative care referrals and effectively managing stress-inducing symptoms. In this field, AI should not replace human shortcomings but assist in their improvement.

Protection and Selection of Data for Algorithm Construction

The use of AI in decision-making requires respect for the autonomy and dignity of individuals. At the same time, it opens possibilities for bias and discrimination in the algorithms used, as these are influenced by training data and reflect existing biases in those data (WHO, 2021b). Some biases that may arise include data bias, where AI algorithms learn from historical data. If these data are biased or reflect existing inequalities, it is likely that the algorithms will reproduce the same biases. For example, if an algorithm is trained using health data that underestimates or discriminates against certain groups of people, especially the elderly given their non-functional status (outside the functional system), it may lead to biased decisions in medical treatment by considering them an unnecessary cost in resource allocation dilemmas.

Another important bias is feature selection, where AI algorithms rely on specific features to make decisions. If these features are correlated with sensitive variables such as ethnicity, gender, or age group, discriminatory biases arise. For example, if an algorithm uses postal addresses as a feature for making hiring decisions, it can result in unintentional geographical or racial discrimination. The design of the algorithm itself can introduce inherent biases. Algorithmic decisions can depend on assumptions or rules that reflect cultural or social prejudices. For instance, a job application classification algorithm may consider certain keywords as indicative of a more suitable candidate, which can lead to indirect discrimination[13].

Another relevant bias is feedback bias, which occurs when AI algorithms learn and adjust based on the responses and feedback they receive. If these responses are biased or discriminatory against certain groups, the algorithm perpetuates and amplifies these biases in negative feedback loops. This has significant impacts in areas such as healthcare, recruitment, the justice system, and other fields where AI decision-making is possible and foreseeable.

There should also be a concern with patient data privacy and security (Nabi, 2018). If these data are not adequately handled, various negative consequences can arise, such as privacy violations, since they contain personal and confidential

---

[13] Authors' notes: You could be fired by an AI, and that may happen soon (Olhar Digital, 2023) (See References).

information, including medical records, diagnoses, past treatments, and other sensitive information. Inadequately protected data can be accessed by unauthorized individuals, resulting in risks like identity theft or being used by criminals to commit fraud, obtain medical services, or engage in other illegal activities in the affected patient's name. Another risk is reputational damage. For example, when there is a security breach or unauthorized disclosure of patient data, healthcare institutions or medical service providers can suffer significant reputational damage, undermining patient and public trust and having long-term repercussions on the confidential relationship between doctor and patient. Furthermore, if the data reveal information about sensitive or stigmatized health conditions, patients may face discrimination in employment, access to insurance, or other life aspects (Bernstein & Pfeiffer, 2023)—an old debate, with significant early contributions noted by various authors in the early 2000s (Sweeney, 2006).

In summary, when health data are processed and analyzed through AI techniques and reveal sensitive information related to medical conditions, negative consequences can arise for affected individuals. This occurs even when there are already discriminatory practices in various sectors based on clients' and patients' work conditions, gender, or illnesses[14]. It is well-known that certain medical conditions are stigmatized in society, meaning that their disclosure can lead to discrimination and prejudice against affected individuals.

The reference to AI implies that health data are processed using algorithms and machine learning techniques to identify patterns, predict risks, or make diagnoses. While AI can be extremely effective in analyzing these datasets, it can also reveal information that people would prefer to keep private. This raises concerns that information revealed by AI could be used in the workplace to discriminate against individuals. For example, a health system could make resource allocation or health promotion decisions for the elderly based on the presence of certain medical conditions, which would be unfair and discriminatory.

Researchers from Google, Microsoft Research, and the FATE (Fairness Transparency Accountability and Ethics in AI) group have reported that large-scale public images can be used to extract sociological information from vulnerable people, creating an "algorithmic bias" and bringing unsuspected ethical implications to projects aiming to gain knowledge from data (Buolamwini

---

[14] (1) In Peru, the National Institute for the Defense of Competition and the Protection of Intellectual Property (Indecopi) imposed fines of over 4 million soles on suppliers from various economic sectors for discrimination against consumers (Indecopi, 2021) (See References).

(2) In the United States, the New Jersey Department of Banking and Insurance ordered insurance companies to provide medical coverage to transgender individuals without discrimination (New Jersey Department of Banking and Insurance, 2023) (See References).

(3) In Chile, it is common for health insurance companies to be fined for 'discriminating' against the entry of new members (La Segunda, 2014) (See References).

& Gebru, 2018). They argue that commercial facial recognition systems exhibit accuracy differences depending on the gender and race of individuals. These disparities can have serious consequences for certain groups, particularly those already disadvantaged or marginalized, such as the elderly (Gebru et al., 2020).

An example of this can be seen in access to protection systems and insurance and how customer data is left unprotected against the incursion of AI, which can influence insurers' decisions in determining premiums or an individual's eligibility for coverage. This results in a situation where people with specific health conditions face difficulties in obtaining affordable insurance or even any coverage at all.

There are many cases of discrimination due to illegal access to personal data, which, in turn, were captured through legal or formal channels of service processes. In the United States, for instance, the insurance company Aetna was fined $1 million in 2019 for violating federal privacy law by disclosing confidential medical information of over 12,000 customers who had HIV or heart disease[15]. The company sent letters to these customers in which the name of the medication they were taking could be seen on the envelope. Additionally, the company had to pay $17 million in compensation to the individuals affected by this incident.

With the technological advancement brought by Artificial Intelligence, it is now possible to use algorithms that learn to assign risk levels for insurability to each applicant based on their medical history, age, gender, among other factors. Furthermore, this also results in requiring applicants of all ages to undergo unnecessary or invasive medical exams.

In the healthcare ecosystem, open practices of data access and disclosure abound, such as allowing employees to access patients' medical data without valid justification and without logging their actions. This implies the risk of data being used for all kinds of illegal and/or unethical purposes, being freely made available to third parties without any control over their destination, such as fraud or manipulation of premiums[16].

Despite many regulatory entities considering these practices a violation of data protection and governance rules, a significant issue arises with data manipulation performed by the intermediaries involved. Many of these rules leave institutions with the ability to control or self-regulate, granting each institution the capacity

---

[15] Authors' notes: Aetna agreed to pay $25.6 million to settle allegations that it disclosed protected health information of thousands of members (BankInfoSecurity, 2023) (See References).

[16] Authors' notes: Several emblematic cases have come to light: In 2019, two former employees of a California hospital were convicted of improperly accessing the records of Sofía Vergara and other celebrities (*NBC News*, 2019). In 2017, a former employee of an Atlanta medical center admitted to selling patient data for tax fraud purposes (*HIPAA Journal*, 2017). In 2015, a former employee of a health insurer leaked VIP patient data in exchange for bribes (*Reuters*, 2015) (See References).

to assess their actions' conditions and decide how to act in the face of these challenges[17].

All this despite the existence of laws such as the California Consumer Privacy Act (CCPA), which prohibits insurers from using personal data to discriminate against clients, or the Health Insurance Portability and Accountability Act (HIPAA) in the United States, which prohibits insurers from using medical data to discriminate against clients in underwriting, renewal, premiums, or granting benefits.

Conclusion

Addressing ethical challenges in the use of AI for the care and monitoring of dependent elderly people requires ensuring that AI is used to guarantee transparency and fairness in the algorithms employed, protecting patient data privacy and security, and promoting open and honest communication with their families. To this end, it is essential to implement strong security measures to protect the data used. This includes encryption, restricting access to confidential information, training staff on data security, adhering to data protection regulations (such as the European Union's General Data Protection Regulation), and conducting periodic audits and checks to identify and address potential data security vulnerabilities.

How AI technology developers will take bioethical dictates into account in constructing these processes remains a question to be better understood. Even more so considering that, generally, the intricacies of decision-making are not well known: Who decides? On what basis? What circumstances lead to one path over another? The presence of a bioethicist, as well as representatives of the elderly in this construction, is necessary, though not sufficient, for the development and refinement of more robust and less biased and manipulated AI tools.

The necessary pragmatism that involves managers' decisions, whether public or private, within a scenario of resource allocation scarcity, has cost-effectiveness as an important reference. If the goal is to improve the quality of life and death, how can this be done without an active interface with Bioethics? Bioethics needs to be invited to participate at some point, or at several points, from the creation to the implementation of predictive AI system models. Its participation is essential for protecting individual subjects who are vulnerated[18], facing collective subjects that will decide their lives, creating significant asymmetrical relationships, impacting their lives and deaths.

---

[17] Authors' note: In Chile, the Financial Market Commission (CMF) issued General Regulation No. 424, which establishes the matters that must be included in the Codes of Conduct to be presented by entities required to self-regulate and that will not participate in the Financial Self-Regulation Committee (Comisión para el Mercado Financiero [CMF], 2023) (See References).

[18] Authors' note: The word vulnerated is used because these patients are no longer merely vulnerable, like every human being is, but have already been afflicted, whether by a terminal condition or by the life conditions to which they are subjected.

At this point, care and the ethics of care play a central role in the formulation and improvement of these mechanisms, which have brought so much expectation and enthusiasm. It is about having care as a kind of leitmotiv, giving it a central and essential dimension in the use of AI:

> "Caring should always take priority over curing for the most obvious of reasons: There is never any certainty that our illnesses can be cured or our death averted. Eventually, they will, and must, triumph. Our victories over sickness and death are always temporary, but our need for support, for caring, in the face of them is always permanent" (Callahan 1990, p. 144).

The beneficial response to the use of AI tools in the care of dependent elderly people will have greater robustness and be in line with solid bioethical principles if these tools are designed to maximize the benefits to the greatest number of people involved in their decisions or minimize harm as much as possible in a scenario where protection, care, respect, and dignity, along with the competence of know-how, are essential parts of a phase of their lives with great restrictions.

References

Avati, A., Jung, K., Harman, S., Downing, L., Ng, A., & Shah, N.H. (2018). Improving palliative care with deep learning. *BMC Medical Informatics and Decision Making*, 18(4), 55–64.

BankInfoSecurity. (2023, January 4). Aetna Settlement: $25.6 Million. https://www.bankinfosecurity.com/aetna-settlement-a-15264

Bloom, D.E., Canning, D., & Fink, G. (2011). Implications of population aging for economic growth. *Oxford Review of Economic Policy*, 26(4), 583–612. https://doi.org/10.1093/oxrep/grq038.

Buolamwini J., & Gebru T. (2018). Gender shades: Intersectional accuracy disparities in commercial gender classification. *Proceedings of the 1st Conference on Fairness, Accountability and Transparency, PMLR*, 81, 77–91, https://proceedings.mlr.press/v81/buolamwini18a.html.

Chen, L.K. (2020). Gerontechnology and artificial intelligence: Better care for older people. *Archives of Gerontolongy and Geriatrics*, 91, 104252. https://doi.org/10.1016/j.archger.2020.104252.

Chen, L.Y., Sumioka, H., Shiomi, M., Ke, L.J., & Chen, L.K. (2020). Effects of teleoperated humanoid robot application in older adults with neurocognitive disorders in Taiwan: A report of three cases. *Aging Medicine and Healthcare*, 11, 67–71. https:// doi.org/10.33879/AMH.2020.066-2001.003.

Callahan, D. (1990). *What Kind of Life: The Limits of Medical Progress*. Georgetown University Press.

Callahan, D. (1995). *Setting Limits: Medical Goals in an Aging Society with a "Response to My Critics"*. Georgetown University Press.

Citro, S. (2010). *Cuerpos Plurales. Antropología de y desde los cuerpos.* Editorial Biblos.

Clusters & Ie. (2023). Pepper, el robot que guía a los pacientes en un hospital belga. https://clustersalud.americaeconomia.com/pepper-el-robot-que-guia-a-los-pacientes-en-un-hospital-belga

Comisión para el Mercado Financiero (CMF). (2023). CMF issued General Standard establishing codes of conduct for entities required to self-regulate. https://www.cmfchile.cl/portal/prensa/615/w3-article-24869.html.

Dolatabadi, E., Zhi, Y.X., Flint, A.J., Mansfield, A., Iaboni, A., & Taati, B. (2019). The feasibility of a vision-based sensor for longitudinal monitoring of mobility in older adults with dementia. *Archives of Gerontology and Geriatrics*, 82, 200–206. https://doi.org/10.1016/j.archger.2019.02.004.

Dufourmantelle, J.A. (2003). *Anne Dufourmantelle convida Jacques Derrida a falar da Hospitalidade*. Editora Escuta.

Elias N. (2001). *A solidão dos moribundos. Envelhecer e morrer*. Jorge Zahar.

Entrepreneur, 2023. https://www.entrepreneur.com/es/tecnologia/rememory-la-plataforma-que-te-permite-hablar-con/442309.

Gebru, T., Raji, I., Mitchell, M., Buolamwini, J., Lee, J., & Denton, E. (2020). Saving Face: Investigating the ethical concerns of facial recognition auditing. *In*: *AIES 2020: Proceedings of the AAAI/ACM Conference on AI, Ethics, and Society*, February (pp. 145–151). https://doi.org/10.1145/3375627.3375820.

Hereafter.AI. (n.d.). https://www.hereafter.ai/

HIPAA Journal. (2017, December 14). Former employee accessed patient data for tax fraud scheme. https://www.hipaajournal.com/former-employee-accessed-patient-data-for-tax-fraud-scheme/.

Huang, C.H., Peng, L.N., & Chen, L.K. (2019). Muscle strength predicting health outcomes of older adults: A narrative review. *Aging Medicine and Healthcare*, 10, 96–99. https://doi.org/10.33879/AMH.2019.1916.

Indecopi. (2021). NP 210712 Sanciones por casos de discriminación. https://repositorio.indecopi.gob.pe/bitstream/handle/11724/8285/NP%20210712%20Sanciones%20por%20casos%20de%20discriminacion.pdf?isAllowed=y&sequence=1.

Intel. (n.d., 2023). *Robotics in Healthcare*. https://www.intel.com/content/www/us/en/healthcare-it/robotics-in-healthcare.html

La Segunda. (2014, July 18). Multan a 4 isapres por discriminar ingreso de afiliados. http://www.lasegunda.com/Noticias/Nacional/2014/07/949513/multan-a-4-isapres-por-discriminar-ingreso-de-afiliados.

Lee, R., & Mason, A. (2011). *Population Aging and the Generational Economy: A Global Perspective*. Edward Elgar Publishing.

Lévinas, E. (2003). *Deus, a Morte e o Tempo*. Almedina. Lewis, A.C.F. (2020). Where bioethics meets machine ethics. *The American Journal of Bioethics*, 20(11), 22–24.

Liu, J.Y., Kor, P.P., Chan, C.P., Kwan, R.Y., & Sze-Ki, D. (2020). The effectiveness of a wearable activity tracker (WAT)-based intervention to improve physical activity levels in sedentary older adults: A systematic review and meta-analysis. *Archives of Gerontology and Geriatrics*, 91, Article 104211. https://doi.org/10.1016/j.archger.2020.104211.

Mannheim, I., Schwartz, E., Xi, W., Buttigeig, S.C., McDonnell-Naughton, M., Wouters, E.J.M., & van Zaalen, Y. (2019). Inclusion of older adults in the research and design of digital technology. *International Journal of Environmental Research and Public Health*, 16, 3718. doi:10.3390/ijerph16193718.Malpani, B. (2020). *Ageism in Artificial Intelligence for Health.* WHO Policy Brief.

Medinaceli, K., & Silva, M. (2021). Impacto y regulación de la Inteligencia Artificial en el ámbito sanitario. *Revista del Instituto de Ciencias Jurídicas de Puebla*, 15(48), 77–113.

Murphree, D.H., Wilson, P.M., Asai, S.W., Quest, D.J., Lin, Y., Mukherjee, P., Chhugani, N., Strand, J.J., Demuth, G., Mead, D., Wright, B., Harrison, A., Soleimani, J., Herasevich, V., Pickering, B.W., & Storlie, C.B. (2021). Improving the delivery of palliative care through predictive modeling and healthcare informatics. *Journal of the American Medical Informatics Association*, 28(6), 1065–1073.

Nabi, J. (2018). How bioethics can shape artificial intelligence and machine learning. *Hastings Center Report*, 48(5), 10–13.

NBC News. (2019, July 25). Two former hospital workers get prison time for peeking at celebrities' records. https://www.nbcnews.com/news/us-news/two-former-hospital-workers-get-prison-time-peeking-celebrities-records-n1023921.

New Jersey Department of Banking and Insurance. (2023, June 28). Comunicado de prensa en español. https://www.state.nj.us/dobi/pressreleases/pr230628espanol.html.

NYT. (2018). Un robot llamado Zora estimula la actividad de los adultos mayores en Francia. The New York Times. https://www.nytimes.com/es/2018/11/30/espanol/zora-robot-ancianos-francia.html.

Nwosu, A.C., Sturgeon, B., McGlinchey, T., Goodwin, C.D.G., Behera, A., Mason, S., Stanley, S., & Payne, T.R. (2019). Robotic technology for palliative and supportive care: Strengths, weaknesses, opportunities, and threats. *Palliative Medicine*, 33(8), 1106–1113.

Olhar Digital. (2023). Você pode ser demitido por uma IA. https://olhardigital.com.br/2023/05/05/pro/voce-podera-ser-demitido-por-uma-ia-e-isso-nao-deve-demorar/

Openmind. (2017). Los robots terapéuticos Paro contribuyen a mejorar la calidad de vida de las personas con demencia y otras afecciones. https://www.bbvaopenmind.com/articulo/robots-terapeuticos-paro-mejoran-calidad-de-vida-dementia

Paro Robots. (2023). Paro Robots. http://www.parorobots.com.

Pessoa F. (1982). *Livro do desassossego* (Vol. II). Ática.

Poblete H. (2018). *El Abrazo de Godzilla. Apuntes Sobre Antropología de la Longevidad.* RAN Editora.

Red Salud. (2020). Hospital Padre Hurtado implements the first robot in Chile to support patient care during Covid-19 times. https://redsalud.ssmso.cl/hospital-padre-hurtado-implementa-el-primer-robot-en-chile-para-apoyar-atencion-de-pacientes-en-tiempos-de-covid-19/

Reuters. (2015, July 2). Ex-Anthem worker accused of allowing data breach for $2 million. https://www.reuters.com/article/us-insurnance-anthem-idUSKCN0PD21B20150702.

Santos, H.G., Zampieri, F.G., Normilio-Silva, K., Silva, G.T., Lima, A.C., Cavalcanti, A.B., & Chiavegatto Filho, A.D.P. (2020). Machine learning to predict 30-day quality-adjusted survival in critically ill patients with câncer. *Journal of Critical Care*, 55, 73–78.

Schramm, F.R. (2010). Existem boas razões para se temer a biotecnociência? *Revista Bioethikos*, 4(2), 189 –197.

Shimada, K., & Tsuneto, S. (2023). Novel method for predicting nonvisible symptoms using machine learning in cancer palliative care. *Scientifics Report*, 13, 12088. https://doi.org/10.1038/s41598-023-39119-0.

Sweeney, L. (2006). Uniquely identifying individuals in the US health insurance system. *Journal of the American Medical Informatics Association*, 13(1), 5 –12.

Thomas, L-V. (1983). *Antropología de la Muerte*. Fondo de Cultura Económica.

Tomašev, N., Glorot, X., Rae, J.W. et al. (2019). Uma abordagem clinicamente aplicável para a previsão contínua de lesão renal aguda futura. *Nature*, 572, 116–119.

United Nations Population Fund. (2014). *Population and Demographic Dynamics*. https://www.unfpa.org/data/world-population/PD

United Nations. (2019). *World Population Prospects 2019*. https://population.un.org/wpp/

Veras, R.P. (2002). *Terceira Idade: Gestão Contemporânea em Saúde*. Rio de Janeiro: Relume-Dumará & Universidade Aberta da Terceira Idade, Universidade do Estado do Rio de Janeiro.

Walzer M. (1983). *Spheres of Justice: A Defense of Pluralism and Equality*. Basic Books.

World Health Organization. (2015). *World Report on Ageing and Health*. https://www.who.int/ageing/events/world-report-2015-launch/en/

World Health Organization. (2021a). *La OMS publica el primer informe mundial sobre inteligencia artificial (IA) aplicada a la salud y seis principios rectores relativos a su concepción y utilización*. https://www.who.int/es/news/item/28-06-2021-who-issues-first-global-report-on-ai-in-health-and-six-guiding-principles-for-its-design-and-use.

World Health Organization. (2021b). *Ageism is a Global Challenge: UN*. 18 March 2021 News release Geneva. https://www.who.int/news/item/18-03-2021-ageism-is-a-global-challenge-un.

World Health Organization. (2022). *Ageism in Artificial Intelligence for Health: WHO Policy Brief*. Geneva: World Health Organization.

Artificial Intelligence Between Totalitarianism and Democracy

Fabiano Tonaco Borges and Armando De Negri Filho

As we conclude this journey through the vast field of artificial intelligence (AI) and bioethics, we are only beginning to understand the profound ethical and social implications of these emerging technologies. This book, *Artificial Intelligence and Bioethics: Perspectives*, brings together a variety of perspectives and analyses that highlight the challenges and opportunities presented by AI.

The recent history of democracies, especially in Latin America and more recently in developed countries, has been marked by advances and setbacks, reflecting a continuous process of construction and destruction. Atílio Borón (2003) reminds us that even the most developed democracies are not immune to regressive attempts that can nullify decades of progress. AI, with its immense transformative potential, can both strengthen and threaten these democratic structures.

Examining the process by which unemployment is generated and employment remuneration worsens in the neoliberal phase of capitalism, while public power withdraws from social protection tasks, it is plausible to consider that the indiscriminate application of AI could exacerbate these conditions, contributing to a "scientific, globalized, and voluntary production of poverty" (Santos, 2000, p. 72). According to Milton Santos, "structural poverty, from a moral and political point of view, is equivalent to a social debt. It is structural and no longer local, nor even national; it becomes globalized, present everywhere in the world. There is a planetary spread and a globalized production of poverty, although it is more present in poor countries" (Santos, 2000, p. 69). The current 'administrative' division of labor, coupled with the deliberate absence of the State in its social regulatory mission, can lead to pervasive and permanent structural poverty, planned in a centralized manner. This scenario not only threatens to increase inequality but also erodes the legitimacy of liberal democracy as people lose trust in a system that cannot guarantee social protection and economic justice.

Under neoliberal theory, the State must favor strong individual rights to private property, the rule of law, and free-market institutions, according to

David Harvey (2013). Sectors previously managed or regulated by the State are transferred to private initiative and deregulated, promoting competition as a primary virtue. Privatization and deregulation, combined with competition, are touted as solutions that eliminate bureaucratic barriers, increase efficiency and productivity, improve quality, and reduce costs. However, this approach often ignores social and environmental consequences, treating labor and the environment as mere commodities

Moreover, neoliberal theorists harbor deep suspicion of democracy, preferring governance by experts and elites. Democratic governance is seen as a potential threat to individual rights and constitutional freedoms, favoring executive and judicial decisions over democratic and parliamentary decision-making. This preference for elitist governance results in a State that tends to favor capital interests over collective well-being, exacerbating inequalities and undermining social solidarity (Harvey, 2013).

Milton Santos offers a critical analysis of this phenomenon by discussing globalization and its implications for democracy and ethics. He argues that the current era is characterized by the extreme use of techniques and norms, leading to an obsessive need for regulations that favor economic centralization and inflexibility of social behaviors. This hegemonic structure, dominated by a technical and market ideology sanctified by science, contributes to the creation of a "technical totalitarianism" that restricts individual and collective freedom.

Neoliberal globalization, as described by Santos (2000), Borón (2003), and Harvey (2013), kills the notion of solidarity and returns man to the primitive condition of "every man for himself". This process is marked by technoscience serving the market, resulting in selective science that benefits specific economic interests to the detriment of general human well-being. The lack of moral progress alongside technical-scientific progress aggravates social inequalities and strengthens the power of economic elites, consolidating a system of exclusion and injustice.

The production of empty signifiers, generated by frustrated democratic demands and potentially enhanced by the false promise of digital inclusion and specially by the use of AI, which ultimately does not alter the condition of class subalternity and therefore the structural inequalities of our globalized societies, and could amplify this frustration of democratic demands into popular demands, easily captured by the discourses of right-wing populism in repeated cycles of accumulated social frustrations (drawing on the categories and terms of analysis of Ernesto Laclau in his work *On Populist Reason*).

The indiscriminate application of AI without ethical considerations can reinforce these dynamics, contributing to the perpetuation of a 'globalitarian' system, as mentioned by Santos. To avoid this, it is essential that AI technologies are developed and implemented with a strong ethical commitment, promoting transparency, justice, and inclusion. The involvement of bioethicists and other ethics specialists in the AI development process is crucial to ensure that these technologies serve the common good, rather than merely hegemonic interests.

Today, both objects and actions derive from technique. "Techniques are everywhere: in production, in circulation, in territory, in politics, in culture. They are also—and permanently—in the body and mind of man. We all live in a tangle of techniques, which in other words means that we are all immersed in the realm of artifice" (Santos, 2000, p. 128). As hegemonic techniques, founded on science and obedient to market imperatives, are now endowed with intentionality, there is also a tendency towards the hegemony of a 'rational' production of things and needs; thus, an exclusive production of other productions, with the multiplication of strictly programmed technical objects that open space for this orgy of things and needs that impose relationships and govern us. "A true tendential totalitarianism of rationality is created—that is, of this hegemonic, dominant rationality—producing, from the respective system, certain things, services, relationships, and ideas" (Santos, 2000, p. 128). This is the primary basis of the production of shortages and scarcities since a considerable portion of society cannot access the things, services, relationships, and ideas that multiply based on hegemonic rationality.

Applying these lessons to the context of AI, we see that technology can be used to reinforce both democratic and authoritarian practices. AI has the potential to increase transparency, equity, and efficiency in democratic processes, but it can also be instrumentalized to enhance surveillance, manipulation, and authoritarian control. The challenge lies in ensuring that the development and implementation of AI are guided by ethical principles that promote freedom, justice, and human dignity.

The post-war period, known as the "Keynesian Recomposition of Capitalism", showed that the interventionist state can play a vital role in promoting social justice and economic stabilization. However, the exhaustion of this golden age led to a reaction against democratic achievements, with proposals to reduce political democracy in favor of capital accumulation. Similarly, the indiscriminate application of AI without ethical considerations can lead to the reinforcement of inequalities and the erosion of democratic freedoms.

The ungovernability described by Borón projects both the destabilizing strategies of those who reject democracy and the aspirations of the large majorities who struggle for decent conditions. If AI is used to sustain an unjust production regime, it can contribute to the delegitimization and destabilization of democracies.

Looking to the future, it is crucial that we continue to question and refine our ethical and regulatory principles regarding AI. This book is just the beginning of an essential conversation about how we can ensure that AI, one of the most powerful technologies of our era, is developed and used in a way that promotes a more just, equitable, and democratic world.

References

Borón, A. (2003). *State, Capitalism, and Democracy in Latin America: In Search of Socialism in the 21st Century*. Buenos Aires: CLACSO.

Harvey, D. (2013). *A Brief History of Neoliberalism*. Oxford: Oxford University Press.
Laclau, E. (2005). *On Populist Reason*. Verso.
Santos, M. (2000). *For Another Globalization: From the Single Thought to Universal Conscience*. Rio de Janeiro: Record.

About the Editors

Luiz Vianna Sobrinho ScM and PhD Post-graduation in Cardiology—Institute of Assistance of Civil Servants of the State of Rio de Janeiro (1989). PhD in Bioethics and Collective Health, by PPGBIOS—National School of Public Health Sérgio Arouca/Fiocruz (2020) and Collaborating Professor at the Department of Human Rights and Health at this school. Author of two books: *Financial Medicine, the Shattered Ethics* (2013) and *Ocaso da clínica—a Medicina de Dados* (2021). Fellow invited scientist (post-doctorate) of the Center for Social Studies of the University of Coimbra (2022). Co-founder of the Latin-America Strategy for Artificial Intelligence.

Leandro Modolo is a medical sociologist. Professor at the Federal Institute of São Paulo (IFSP). PhD in Social Sciences from the State University of São Paulo (UNESP), Brazil. Attend doctoral fellowship at Collective Health from State University of Campinas (UNICAMP) in São Paulo, Brazil, and participates in a scholarship at FIOCRUZ Strategy for the 2030 Agenda, sustainable development and the structuring of the Strategic Health Prospecting and Management Network. He is a member of the Nucleus of Studies in Economics, Technology and Society (NETS) of the Federal University of Ceará (UFCE), participates in the Conexões Group (Unicamp), and the Occupational Health Study Laboratory—Ester (Unicamp). Co-founder of the Latin-America Strategy for Artificial Intelligence.

Maíra Araújo de Santana is a biomedical engineer from the Federal University of Pernambuco (UFPE), Brazil (2017) with experience in the United States, where she attended an academic year at the University of Alabama at Birmingham, and worked as a researcher at Duke University, at the Carl E. Ravin Advanced Imaging Laboratories (RAI Labs), developing specific knowledge in the medical image processing field. She holds a Master's degree in Biomedical Engineering (2020), and a PhD in Computer Engineering (2023). She currently works as a postdoctoral researcher and on the faculty of the Specialization in Data Science and Digital Health at the Department of Biomedical Engineering of UFPE. She is a member of the Biomedical Computing Research Group (UFPE) and develops research with artificial intelligence applied to health, in areas such as pattern recognition for early diagnosis of breast cancer, Affective Computing, and Applied Neuroscience.

Giselle Machado Magalhães Moreno Co-founder of the Latin-America Strategy for Artificial Intelligence. She has a Master's degree in Neuropsychiatry and Behavioral Sciences (Federal University of Pernambuco, 2012–2014) and PhD in Biochemistry and Physiology (Federal University of Pernambuco, 2014–2018). She is linked, as a postdoctoral researcher, at the INterdisciplinary CLimate INvestigation cEnter (INCLINE) at University of São Paulo, Brazil. Her research interests include public health, digital epidemiology, neuroscience, and artificial intelligence applied to medicine and biology.

Fabiano Tonaco Borges Co-founder of the Latin-American Strategy for Artificial Intelligence. Professor at the Collective Health Institute of the Federal University of Mato Grosso (UFMT), Researcher of the Biomedical Computing Laboratory of the Federal University of Pernambuco (UFPE), and Consultant of the Brazilian Ministry of Health/Pan-Americam Health Organization. Oral Surgeon by training. Holds a master's and a doctorate degree in Social and Preventive Dentistry from UNESP in Araçatuba, São Paulo, Brazil, with a sandwich scholarship at the University of Massachusetts Lowell (2010–2011). Completed a post-doctoral fellowship at UNICAMP in Campinas, São Paulo, Brazil, in association with the Andalusian School of Public Health in Granada, Spain (2014–2015), and was an invited post-doctoral fellow at the University of Tübingen, in Tübingen, Baden-Württemberg, Germany (2019–2020). His research work focuses on health policy analysis and Artificial Intelligence.

Wellington Pinheiro dos Santos holds a degree in Electrical and Electronic Engineering (2001) and a Master's degree in Electrical Engineering (2003) from the Federal University of Pernambuco, and a PhD in Electrical Engineering from the Federal University of Campina Grande (2009). He is currently an Associate Professor (exclusive dedication) at the Department of Biomedical Engineering at the Center for Technology and Geosciences/School of Engineering of Pernambuco, Federal University of Pernambuco, working in the Undergraduate Program in Biomedical Engineering and the Graduate Program in Biomedical Engineering. He is also a member of the Graduate Program in Computer Engineering at Escola Politécnica de Pernambuco, Universidade de Pernambuco, since 2009. He has experience in the field of Computer Science, with an emphasis on Graphics Processing, working mainly in the following areas: topics: digital image processing, pattern recognition, computer vision, evolutionary computing, numerical optimization methods, computational intelligence, image formation techniques, virtual reality, game design, and applications of Computing and Engineering in Medicine and Biology. He is a member of the Brazilian Society of Biomedical Engineering (SBEB), of the Brazilian Society of Computational Intelligence (SBIC, ex-SBRN), and of the International Federation of Medical and Biological Engineering (IFMBE).

About the Authors

Armando De Negri Filho born in Porto Alegre, Brazil, is a renowned Brazilian-Italian specialist in public health and health policy. With an impressive academic background, he graduated in Medicine from the Federal University of Rio Grande do Sul in 1990. De Negri holds a PhD in Preventive Medicine with a focus on Health Policy and Systems (2016), in addition to multiple master's degrees: Clinical Management and Medical Coordination from the Carlos III Institute, Madrid, Spain (2015), Public Health with an emphasis on Global Health and Health Diplomacy from the National School of Public Health—FIOCRUZ, Rio de Janeiro (2014), and Epidemiology from the Federal University of Pelotas (1995). He is also a specialist in Emergency Medicine certified by ABRAMEDE/AMB Brazil (2018).

In his professional career, De Negri is currently an International Specialist and Advisor on Health Systems and Services of the Pan American Health Organization/World Health Organization at the Mexico City Office since December 2021. Since March 2021, he has been a Senior Researcher at the Center for International Relations in Health (CRIS) at FIOCRUZ, Rio de Janeiro. Between 2020 and 2021, he was an independent specialist member of the Expert Mechanism on the Right to Development of the United Nations Human Rights Council and was shortlisted for the position of Special Rapporteur on the Right to Health at the same council in 2014.

De Negri is the founder and coordinator of the Innovations Laboratory on Health Planning, Management, Evaluation, and Regulation of Health Policies, Systems, Networks, and Services (LIGRESS) at the Heart Hospital (HCor) in São Paulo, where he served from 2009 to 2017. Additionally, he has served as an advisor on health and social policies for the governments of Venezuela, Bogotá (Colombia), and Paraguay, and as an advisor to the Equity Observatory of the Council for Economic and Social Development of the Presidency of the Republic of Brazil (2006–2009). In Porto Alegre, he held the position of Director of Health Planning and Strategic Projects for the city on two occasions (1993–1997, 2001–2003). He was also the General Coordinator of the Brazilian Cooperation Network on Emergencies (RBCE) for several periods, from 1995 to 2021.

Carlos Eduardo Gomes Siqueira is a retired Professor Emeritus in the Department of Urban Planning and Community Development at the School for the Environment at the University of Massachusetts Boston, and Coordinator of the Transnational Brazilian Project at the Mauricio Gastón Institute of Latino

Community Development and Public Policy. Dr. Siqueira graduated in Medicine at the Federal University of Rio de Janeiro, Brazil. He holds a Doctorate in Work Environment Policy from UMass Lowell and a Master in Public Health from the Johns Hopkins Bloomberg School of Public Health. He was a Fulbright Scholar at the Graduate Entry Medical School at the University of Limerick, Ireland in 2019. Over his career he has researched work environment policy issues, environmental justice for Brazilian immigrants, Brazilian health policy, and health and safety disparities at work in the U.S. and Latin America. He authored many book chapters in English or Portuguese, a few books, and dozens of peer-reviewed articles on public health issues.

Raphael Augusto Teixeira de Aguiar Adjunct Professor in the Department of Preventive and Social Medicine at the Medical School of the Federal University of Minas Gerais (UFMG), where he also served as head from 2019 to 2023. Current vice-director of the Center for Collective Health Education (NESCON/UFMG). Raphael holds a degree in Medicine (1999), a Master's in Public Health (2003), and a PhD in Education (2010), all from UFMG. Formerly a public health physician at the UFMG Clinical Hospital, he has also been a consultant for the Brazilian Ministry of Health and the coordinator of educational management at the Open University of SUS (UNA-SUS). Currently pursuing a specialization in Data Science and Big Data at the Pontifical Catholic University of Minas Gerais.

Karla Figueiredo received a BSc degree in Electrical Engineering from the Federal University of Rio de Janeiro (UFRJ), Brazil in 1990, an MSc degree in Electrical Engineering from the Pontifical Catholic University of Rio de Janeiro (PUC-Rio), Brazil, in 1994, and the PhD in Computer Systems from the Pontifical Catholic University of Rio de Janeiro (PUC-Rio), Brazil, in 2003. She is currently an Associate Professor of the Department of Informatics and Computer Science/Institute of Statistics and Mathematics at Rio de Janeiro University State, working in the Undergraduate and Postgraduate Program in Telemedicine and Telehealth. She is also a member of the Graduate Program in Computer Science at the Institute of Statistics and Mathematics at Rio de Janeiro University State. She has been working or managing more than 45 projects in Artificial Intelligence, over the last 29 years. She is the author of 10 book chapters and more than 100 scientific articles on Soft Computing and Machine Learning. She has supervised more than 35 doctoral theses and master's dissertations. Karla Figueiredo also coordinates the LIA2 laboratories (Applied Artificial Intelligence laboratory) at UERJ and the Applied Intelligence and Robotics Laboratory at PUC-Rio. She is also vice-coordinator of the Artificial Intelligence Center of Rio de Janeiro (CIA-Rio). She is a member of the Institute of Electrical and Electronics Engineers (IEEE), the Association for Computing Machinery (ACM), Brazilian Society of Computational and the Brazilian Society of Computational Intelligence (SBIC).

Ciro Augusto Floriani ScM and PhD, Professor, Faculty of Medicine at the University Estácio de Sá – Idomed, Campus Vista Carioca, Rio de Janeiro, Brazil.

Hernán Poblete Miranda Social Anthropologist from the Bolivarian University of Santiago, Chile, with undergraduate studies in Design and postgraduate studies in Complex Social Systems from the University of Chile. Currently an independent researcher at The Machine Learning Solutions, a private research laboratory developing applied artificial intelligence solutions in various fields, such as longevity, education, and health. Additionally, he has developed applications in financial services, customer relationship management (CRM), and data analytics, leveraging socio-anthropological insights and machine learning techniques to enhance decision-making processes and operational efficiencies. Among his publications are the book *Composite Commercial Microcenters: Economic Anthropology* (2018), which outlines an ethnographic research method to describe economic processes in favelas and low-income neighborhoods; *El Abrazo de Godzilla* (2021), where he proposes a methodological approach and the construction of an anthropology of longevity as opposed to the concept of 'aging'; and the white paper "Centralidad en el Usuario Operada con Inteligencia Artificial" (2022), where he analyzes and advocates for the use of social anthropology in designing narratives for automatic language modeling. Co-founder of ELA-IA (Latin American Artificial Intelligence Strategy).

Fermin Roland Schramm born in Zurich, Switzerland, in 1948. Public health specialist and bioethicist, graduated in Literature at the University of Geneva, Switzerland, in 1974; with postgraduate studies in semiotics at the School for Advanced Studies in the Social Sciences (EHESS) from 1975 to 1978; doctorate in Public Health at the National School of Public Health (ENSP) of the Oswaldo Cruz Foundation (1993) and postdoctoral studies in bioethics at the University of Chile; teaching officer at the University of Geneva (1974–1983); researcher and professor of philosophy of sciences and bioethics at ENSP since 1994.

Priscilla Matsushita Lawyer specialized in bioethics and medical ethics from the Faculty of Law of the University of Lisbon, Portugal (FDUL) and in Data Protection and Health Compliance from PUC – Campinas, in São Paulo. Works as a data protection and privacy consultant, certified by IAPP, as a CDPO (Certified Data Protection Officer) and CIPM (Certified Information Privacy Manager). Professor of privacy and personal data protection. Co-founder and General-Secretary of ELA-IA (Latin American Artificial Intelligence Strategy).

Crístyny Almeida Lawyer and Compliance Consultant. Postgraduate in Administrative and Public Law. Postgraduate in Tax Law. Active Member of the Permanent Study Commission on Compliance of OAB/MT. Co-founder of ELA-IA (Latin American Artificial Intelligence Strategy).

Index

T

U

W

For Product Safety Concerns and Information please contact our
EU representative GPSR@taylorandfrancis.com Taylor & Francis
Verlag GmbH, Kaufingerstraße 24, 80331 München, Germany